# STATE, FEDERAL, AND C.I.T.E.S. REGULATIONS FOR HERPETOLOGISTS

by Norman Frank and Erica Ramus

Cover photograph — Wood Turtle *(Clemmys insculpta)*

ISBN # 0-9641032-1-4
© 1994 *Reptile & Amphibian Magazine*
Published by Reptile & Amphibian Magazine
A Division of N G Publishing Inc. Pottsville, Pa.

No portion of this book may be reproduced in any form, nor transmitted or stored in any database, electronic or computer system, or information retrieval network for public or private use without the written consent of the copyright holder and publisher.

# AUTHORS' STATEMENT

One of the most confusing and least understood areas in herpetology is the regulatory scene. While many reptile and amphibian enthusiasts enjoy collecting and keeping herps, they may not be aware of the various rules and regulations that apply to their hobby.

We contacted the wildlife regulatory agencies for each state in the U.S. (plus Puerto Rico and the Virgin Islands) and requested information on state reptile and amphibian species and also all rules and regulations that apply to persons collecting and possessing herps in their state. Most responded immediately and provided us with more information than we could possibly print here. While some states do not have their own Endangered/Threatened lists for herps, they do follow the Federal list. For each state (listed alphabetically) we listed the wildlife regulatory agency's name, address, and phone number, Endangered/Threatened species, Species of Special Concern, and then a summary of the state's rules and regulations.

Keep in mind that wildlife regulations are not written in stone; the statuses of native species are continuously being updated, and regulations can change frequently. What follows is only a *summary* of each state's regulations, and is not meant to be a comprehensive rule-book. Check with your state agency for a complete list of their collecting regulations. Also, while many states don't regulate keeping non-native herps, local governments may have such rules, so "exotic" pets may be covered under city laws.

Wildlife is protected on two fronts: internationally (through C.I.T.E.S.), and nationally (through the Endangered Species Act). The agreement to promote wildlife conservation internationally, conceived in 1975 as a result of The Convention on International Trade in Endangered Species of Wild Fauna and Flora, is commonly referred to as C.I.T.E.S.. Its purpose is to protect wild plants and animals; most wildlife importing and exporting nations are signatories.

C.I.T.E.S. lists certain species in three appendices. Appendix I contains over 600 life forms which are threatened with extinction. Member nations agree to ban commercial trade in these animals and plants. Appendix II wildlife are not considered threatened with extinction at present, but may become so if trade is not controlled. Thus, while commerce is allowed, permits from both the exporting and importing nations must be obtained. There are over 2,300 animals and 24,000 plants in Appendix II. In addition, "look-alikes" that are not themselves threatened, but resemble Appendix I and II individuals, are governed to help enforcement agents. For example, all crocodilians and boas not listed on Appendix I and II are covered so that custom agents know shipments containing these animals must be checked closely. C.I.T.E.S. also publishes a third appendix; this allows member nations to control export of native species which are locally depleted. Appendix III plants and animals may be traded if the country of origin issues a permit.

C.I.T.E.S. is a political agreement, and individual countries are responsible for local enforcement. Members must adhere to the minimum C.I.T.E.S. standards, although their laws may be stricter, if desired. For example, many nations, such as Venezuela and Brazil, forbid export of all native species, whether they are given C.I.T.E.S. protection or not. Other countries, however, lag far behind in C.I.T.E.S. enforcement. Japan, for example, signed as a party to the agreement in 1980 but still deals in whale and sea turtle products; both groups are listed on Appendix I. Party nations agree to submit annual reports on trade and attend biennial conferences. They agree to confiscate smuggled goods and penalize violators. C.I.T.E.S. is financed by member countries according to a contribution scale drawn up by the United Nations. Critics complain, maybe justifiably, that only about half of all parties maintain acceptable reporting and enforcement of C.I.T.E.S. regulations. Still, it is the most widely accepted conservation program in the world. The U.S. and Canada are member nations, both having signed in July, 1975.

The U.S. government also keeps its own list of "Endangered and Threatened Wildlife and Plants" worldwide (commonly abbreviated the "T&E list"). Under the Endangered Species Act of 1973, the Department of the Interior and U.S. Fish & Wildlife Service protect plants and animals which are on the Federal list, and implements C.I.T.E.S. regulations within the U.S. Species classified as C.I.T.E.S. Appendix I, II, or III are not permitted to be imported or exported without the proper C.I.T.E.S. permits. In addition, the Lacey Act makes it illegal to import any animal taken in violation of foreign law.

While many states have their own T&E state lists, those that do not rely heavily on the Federal list to decide which species in their state need to be protected and regulated. The C.I.T.E.S. and Federal lists have been combined here, and species are grouped taxonomically.

# ALABAMA

Department of Conservation and Natural Resources
64 North Union St.
P.O. Box 301456
Montgomery, AL 36130-1456
Phone: (205) 242-3469

**ENDANGERED:**
  Leatherback Sea Turtle *(Dermochelys coriacea)*
  Hawksbill Sea Turtle *(Eretmochelys imbricata)*
  Kemp's Ridley Sea Turtle *(Lepidochelys kempii)*
  Alabama Redbelly Turtle *(Pseudemys alabamensis)*

**THREATENED:**
  American Alligator *(Alligator mississippiensis)*
  Eastern Indigo Snake *(Drymarchon corais couperi)*
  Loggerhead Sea Turtle *(Caretta caretta)*
  Green Sea Turtle *(Chelonia mydas)*
  Gopher Tortoise *(Gopherus polyphemus)*
  Flattened Musk Turtle *(Sternotherus depressus)*
  Red Hills Salamander *(Phaeognathus hubrichti)*

**REGULATIONS:**
It is unlawful to take, capture, kill, or attempt to take, capture, kill; possess, sell, trade for anything of monetary value, or offer to sell or trade for anything of monetary value, the following nongame wildlife species (or any parts or reproductive products of such species) without a scientific collection permit or written permission from the Commissioner, Department of Conservation and Natural Resources, which shall state what the permittee may do with regard to said species:

| | |
|---|---|
| Dusky Gopher Frog | Eastern Indigo Snake |
| Pine Barrens Treefrog | Florida Pine Snake |
| Eastern Hellbender | Gulf Salt Marsh Snake |
| Flatwoods Salamander | Southern Hognose Snake |
| Green Salamander | Mississippi Diamondback Terrapin |
| Red Hills Salamander | Gopher Tortoise |
| Seal Salamander | Alabama Map Turtle |
| Tennessee Cave Salamander | Alabama Redbelly Turtle |
| Eastern Coachwhip | Alligator Snapping Turtle |
| Black Pine Snake | Barbour's Map Turtle |

# ALASKA

Department of Fish and Game
Division of Sport Fish
P.O. Box 25526
Juneau, AK 99802-5526
Phone: (907) 465-4180
Fax: (907) 465-2772

**ENDANGERED:**
    Under the state's Endangered Species Program, there are no reptiles or amphibians listed as Endangered. Four marine turtles occur in Alaskan waters, although rarely, and these are covered under the Federal Endangered Species Act—the Green Sea Turtle, the Loggerhead Sea Turtle, the Olive Ridley Sea Turtle, and the Leatherback Sea Turtle.

**THREATENED:**
    Alaska has no "threatened" category.

**SPECIES OF SPECIAL CONCERN:**
    No reptiles or amphibians are listed as species of special concern.

**REGULATIONS:**
    Alaska has no regulations that specifically provide for any type of harvest or commercialization of amphibians. It is illegal to collect, kill, or retail any live amphibians for recreational purposes.
    However, amphibians are included in the state's definition of fish, and the Department of Fish and Game has the authority to issue collecting permits for scientific and educational collections of amphibians. The state has very few requests to collect amphibians, however, during the last several years, there has been interest in learning more about the distribution of the Spotted Frog in southeast Alaska, and the state has issued permits for their collection.

**NOTES:**
    There are very few amphibians and reptiles in Alaska. The only terrestrial reptile that has been documented in the state is the Garter Snake, and it is so rare that it could probably be considered almost an "accidental" type species. There are seven species of amphibians that occur in Alaska. One, the Pacific Treefrog, is not native to the state, but became established some years ago after a release into Ward Lake, near Ketchikan. The Wood Frog is probably the most widely occurring species and southeast Alaska is home to more species than in the more northerly parts of the state.

# *ARIZONA*

Game and Fish Department
2221 W. Greenway Rd.
Phoenix, AZ 85023
Phone: (602) 942-3000

**ENDANGERED:**
    Arizona Skink *(Eumeces gilberti arizonensis)*
    Massasauga *(Sistrurus catenatus)*
    Huachuca Tiger Salamander *(Ambystoma tigrinum stebbinsi)*
    Barking Frog *(Hylactophryne [Eleutherodactylus] augusti)*
    Tarahumara Frog *(Rana tarahumarae)*
    Plains Leopard Frog *(Rana blairi)*

**THREATENED:**
    Flat-tailed Horned Lizard *(Phrynosoma mcallii)*
    Chiricahua Leopard Frog *(Rana chiricahuensis)*

**CANDIDATE SPECIES:**
    Desert Tortoise *(Gopherus agassizii)*
    Colorado Desert Fringe-toed Lizard *(Uma notata)*
    Mohave Fringe-toed Lizard *(Uma scoparia)*
    Mexican Garter Snake *(Thamnophis eques)*
    Narrow-headed Garter Snake *(Thamnophis rufipunctatus)*
    Brown Vine Snake *(Oxybelis aeneus)*
    Arizona Ridge-nosed Rattlesnake *(Crotalus willardi willardi)*
    Northern Casque-headed Frog *(Pternohyla fodiens)*
    Northern Leopard Frog *(Rana pipiens)*
    Lowland Leopard Frog *(Rana yavapaiensis)*
    Great Plains Narrow-mouthed Toad *(Gastrophryne olivacea)*

**REGULATIONS:**
    A valid hunting license is required for taking reptiles other than softshell turtles. A valid fishing license is required for taking any aquatic wildlife, including amphibians or softshell turtles, from public waters. The following species are "Restricted Live Wildlife" and a special license or an exemption is required in order to engage in any activity related to these animals: Order Crocodylia—caimans, gavials, crocodiles, alligators; Family Chelydridae—snapping turtles; *Gopherus* and *Xerobates* species—gopher tortoises; Family Helodermatidae—Gila Monster, Beaded Lizard; Family Elapidae—cobras, mambas, coral snakes, kraits, Australian elapids; Family Viperidae—true vipers and pit vipers, including rattlesnakes; Family Atractaspidae—burrowing asps; *Dispholidus typus*—Boomslang; *Thelotornis kirtlandii*—Bird Snake; *Rhabdophis* species—keelbacks; *Xenopus* species—clawed frogs; *Bufo horribilis, B. marinus, B. paracnemis*—giant toads.
    For more information, Arizona has a detailed pamphlet available: *Arizona Reptile & Amphibian Regulations 1994.*

# ARKANSAS

Game and Fish Commission
2 Natural Resources Dr.
Little Rock, AR 72205
Phone: (501) 223-6300

**ENDANGERED:**
No herptiles are currently on the Arkansas Endangered Species List. The American Alligator is on the Federal List for Arkansas, and cannot be taken.

**SPECIES OF SPECIAL CONCERN:**

- Buttermilk Racer
- Western Diamondback Rattlesnake
- Corn Snake
- Dusty Hognose Snake
- Louisiana Milk Snake
- Texas Coral Snake
- Green Water Snake
- Graham's Crayfish Snake
- Gulf Crayfish Snake
- Queen Snake
- Ground Snake
- Western Plains Garter Snake
- Western Chicken Turtle
- Razorback Musk Turtle
- Ornate Box Turtle
- Southeastern Five-lined Skink
- Great Plains Skink
- Southern Prairie Skink
- Mole Salamander
- Ozark Hellbender
- Spotted Dusky Salamander
- Oklahoma Salamander
- Four-toed Salamander
- Caddo Mountain Salamander
- Fourche Mountain Salamander
- Rich Mountain Salamander
- Southern Redback Salamander
- Ringed Salamander
- Bird-voiced Treefrog
- Illinois Chorus Frog
- Strecker's Chorus Frog
- Western Chorus Frog
- Northern Crawfish Frog
- Plains Spadefoot Toad
- Eastern Spadefoot Toad
- Hurter's Spadefoot Toad

**REGULATIONS:**
In 1987 the Arkansas Game and Fish Commission incorporated various revisions in the AGFC Official Codebook of Regulations which prohibit the taking of nongame animals from the wild for commercial purposes. In addition, a new provision pertains to a Commercial Nongame Breeder's Permit established by the Commission. Under authorization of this permit, persons in present possession of nongame wildlife may now market such animals or their offspring. Also, animals obtained by legal means, not from the wild of the state, may be bought, sold, or used as broodstock. Currently, six parties are permitted as Commercial Nongame Breeders in Arkansas.

Several species of salamanders endemic to Arkansas are suffering from overcollection by both legal and illegal collectors. Several of these are currently under study to determine their status as pertains to the Endangered Species Act of 1973. The following species may not be collected for any reason without special permission from the state's Endangered, Nongame, and Urban Wildlife Office: Fourche Mountain Salamander, Caddo Mountain Salamander, Southern Redback Salamander, Ozark Hellbender, Ringed Salamander, Mole Salamander, Spotted Dusky Salamander, Oklahoma Salamander, and the Four-toed Salamander.

# CALIFORNIA

Department of Fish and Game
Wildlife Protection Division
1416 Ninth St.
Sacramento, CA 95814
Phone: (916) 653-1725

**PROTECTED:**

- Santa Cruz Long-toed Salamander
- Siskiyou Mountain Salamander
- Desert Slender Salamander
- Kern Canyon Slender Salamander
- Tehachapi Slender Salamander
- Shasta Salamander
- Limestone Salamander
- Mt. Lyell Salamander
- Yellow-blotched Salamander
- Inyo Mountains Salamander
- Black Toad
- Red-legged Frog
- Southwestern Toad
- Coachella Valley Fringe-toed Lizard
- Blunt-nosed Leopard Lizard
- Panamint Alligator Lizard
- Island Night Lizard
- Flat-tailed Horned Lizard
- Switak's Barefoot Gecko
- Leaf-toed Gecko
- Granite Night Lizard
- Orange-throated Whiptail
- Black Legless Lizard
- San Diego Horned Lizard
- Banded Gila Monster
- Southern Rubber Boa
- San Francisco Garter Snake
- Alameda Striped Garter Snake
- Giant Garter Snake
- San Diego Mountain Kingsnake
- Desert Tortoise
- Sonora Mud Turtle
- Yellow Mud Turtle

**REGULATIONS:**

None of the above protected species may be taken or possessed at any time, except under special permit. A California sport fishing license is required for taking amphibians and reptiles, except rattlesnakes. Researchers, students, and teachers interested in collecting California wildlife species for scientific, educational, or breeding purposes must obtain a Scientific Collection Permit from the Department of Fish and Game.

Persons wishing to collect native herps and breed them in captivity for commercial purposes must obtain a Native Reptile and Amphibian Captive Propagation Permit. Permit holders must maintain detailed records, including accounts of all adults, dates eggs laid, number of eggs laid, number of births, birth dates, and accounts and dates of sales and purchases of offspring. Permittees are expressly prohibited from releasing stocks or offspring to the wild. It is illegal to release into the wilds of California any animal which is not native to California.

The following species are prohibited and it is unlawful to import, transport, or possess these species alive except under permit: *Bufo marinus, B. paracnemis, B. horribilis* (giant toads) and all other large toads from Mexico, Central and South America; all frogs of the genus *Xenopus*; all crocodiles, caimans, alligators, and gavials; snapping turtles; all snakes of the Family Elapidae (cobras, coral snakes, mambas, kraits, etc.); all snakes of the Family Viperidae (adders and vipers); all Crotalidae—except *Crotalus viridis, C. atrox, C. ruber, C. scutulatus, C. mitchelli,* and *C. cerastes; Dispholidus typus; Theoltornis kirlandii; Heloderma suspectum suspectum.*

# COLORADO

Department of Natural Resources
Division of Wildlife
6060 Broadway
Denver, CO 80216-1000
Phone: (303) 297-1192


**ENDANGERED:**
    Midget Faded Rattlesnake *(Crotalus viridis concolor)*
    Massasauga *(Sistrurus catenatus)*
    Western Toad *(Bufo boreas)*

**THREATENED:**
    Wood Frog *(Rana sylvatica)*

**REGULATIONS:**
    Up to six each of species of reptiles and amphibians (except Endangered or Threatened species) may be kept in captivity.

    A fishing license is required for all persons 15 years of age or older who collect frogs in Colorado. Persons collecting amphibians for bait, commercial sale, or personal use by fishing, cast net, dip net, net, trap, or seine must have in possession one of the following: a valid fishing license if collecting for personal use, or a valid commercial license if collecting for commercial purposes.

    The taking, possession, and use of Bullfrogs and the aquatic gilled form of the Tiger Salamander for private and commercial use is permitted. Bullfrog open season is year-round; the daily bag and possession limit is 10 frogs. They may be taken by fishing, archery, hand, and by the use of gigs and nets. Artificial light may be used while frogging. The gilled form of the aquatic Tiger Salamander open season is year-round; daily bag and possession limit is 20 animals less than five inches in length. The possession of terrestrial adult (land form) Tiger Salamanders is limited to six. Salamander larvae may be taken by fishing, by hand, traps, and by the use of seines and nets. The taking of Boreal Toads is prohibited.

# CONNECTICUT

Department of Environmental Protection
Wildlife Division
79 Elm St.
Hartford, CT 06106
Phone: (203) 566-4683

**ENDANGERED:**
    Bog Turtle *(Clemmys muhlenbergii)*
    Leatherback Sea Turtle *(Dermochelys coriacea)*
    Atlantic Ridley Sea Turtle *(Lepidochelys kempii)*
    Timber Rattlesnake *(Crotalus horridus)*
    Eastern Spadefoot Toad *(Scaphiopus holbrookii)*

**THREATENED:**
    Loggerhead Sea Turtle *(Caretta caretta)*
    Atlantic Green Sea Turtle *(Chelonia mydas)*
    Five-lined Skink *(Eumeces fasciatus)*
    Blue-spotted Salamander, diploid populations *(Ambystoma laterale)*
    Northern Spring Salamander *(Gyrinophilus porphyriticus)*
    Northern Slimy Salamander *(Plethodon glutinosus)*

**SPECIES OF SPECIAL CONCERN:**
    Eastern Hognose Snake *(Heterodon platirhinos)*
    Eastern Ribbon Snake *(Thamnophis sauritus)*
    Jefferson Salamander "complex" *(Ambystoma jeffersonianum)*
    Blue-spotted Salamander "complex" *(Ambystoma laterale)*

**REGULATIONS:**
    At any time, no person shall possess:
        —in excess of 3 Spotted Salamanders
        —in excess of 3 Marbled Salamanders
        —any Wood Turtle
        —in excess of 1 Eastern Box Turtle
    Open seasons:
        Spotted Salamanders, May 1—August 31
        Marbled Salamanders, May 1—August 31
        Bog Turtles and Wood Turtles, none
        Eastern Box Turtles, July 1—August 31
        Black Rat Snakes, May 1—August 31
        Diamondback Terrapins, August 1—April 30
    Any person owning or keeping a Bog Turtle shall submit the following information to the Wildlife Bureau: name and address of owner; name and address of keeper; type and number of turtles owned or kept; date each turtle was acquired; approximate age of each turtle; address at which each turtle is kept; and any other information which the Commissioner deems necessary.

# DELAWARE

Department of Natural Resources
Division of Fish & Wildlife
P.O. Box 1401
Dover, DE 19903
Phone: (302) 739-4782

**ENDANGERED:**
- Hawksbill Sea Turtle (*Eretmochelys imbricata*)
- Leatherback Sea Turtle *(Dermochelys coriacea)*
- Kemp's Ridley Sea Turtle *(Lepidochelys kempii)*
- Tiger Salamander *(Ambystoma tigrinum)*
- Bog Turtle *(Clemmys muhlenbergii)*
- Cope's Gray Treefrog *(Hyla chrysocelis)*
- Barking Treefrog *(Hyla gratiosa)*

**THREATENED:**
- Green Sea Turtle *(Chelonia mydas)*
- Loggerhead Sea Turtle *(Caretta caretta)*

**REGULATIONS:**

The Delaware Division of Fish & Wildlife has been engaged in the complex process of revising the State Wildlife Statutes. Included in the proposed revisions are sections dealing with listing and protection. The current list is badly in need of rigorous review, and likely does not entirely encompass all qualifying species. Currently, the coordinator knows of several strong candidates for listing. So far, they have awaited the code revision; however, some situations may be serious enough for remedial efforts.

If you have further questions about listed wildlife species, listing processes, or reviews pertaining to proposed projects, contact the coordinator of the Nongame & Endangered Species Program at the above phone number. Requests regarding specific sites also should be accompanied by a letter describing the reason for and nature of your request and a map clearly delineating the site.

# *FLORIDA*

Game and Fresh Water Fish Commission
Little River Ranch, Route 7, Box 3055
Quincy, FL 32351
Phone: (904) 627-9674
  850

**ENDANGERED:**
    American Crocodile *(Crocodylus acutus)*
    Atlantic Green Sea Turtle *(Chelonia mydas mydas)*
    Atlantic Hawksbill Turtle *(Eretmochelys imbricata imbricata)*
    Atlantic Ridley Sea Turtle *(Lepidochelys kempii)*
    Leatherback Sea Turtle *(Dermochelys coriacea)*
    Mud Turtle *(Kinosternon bauri)*—Lower Keys population only

**THREATENED:**
    Loggerhead Sea Turtle *(Caretta caretta)*
    Blue-tailed Mole Skink *(Eumeces egregius lividus)*
    Sand Skink *(Neoseps reynoldsi)*
    Big Pine Key Ringneck Snake *(Diadophis punctatus acricus)*
    Miami Blackheaded Snake *(Tantilla oolitica)*
    Short-tailed Snake *(Stilosoma extenuatum)*
    Florida Brown Snake *(Storeria dekayi victa)*—Lower Keys population only
    Indigo Snake *(Drymarchon corais)*
    Atlantic Salt Marsh Water Snake *(Nerodia fasciata taeniata)*

**SPECIES OF SPECIAL CONCERN:**
    Gopher Frog *(Rana areolata)*
    Pine Barrens Treefrog *(Hyla andersonii)*
    Florida Bog Frog *(Rana okalossae)*
    Georgia Blind Salamander *(Haideotriton wallacei)*
    Alligator Snapping Turtle *(Macrochelys temmincki)*
    Suwannee Cooter *(Chrysemys concinna suwanniensis)*
    Barbour's Map Turtle *(Graptemys barbouri)*
    Gopher Tortoise *(Gopherus polyphemus)*
    American Alligator *(Alligator mississippiensis)*
    Florida Key Mole Skink *(Eumeces egregius egregius)*
    Red Rat Snake *(Elaphe guttata guttata)*
    Florida Pine Snake *(Pituophis melanoleucus megitus)*

**REGULATIONS:**
    Any person who wishes to exhibit live reptiles to the public or to possess for sale or sell live reptiles (except anoles) must obtain a License to Possess Wildlife for Exhibition or Public Sale. A Venomous Reptile Permit is required by anyone wishing to keep, possess, or exhibit any venomous reptile. It is illegal to buy, sell, or possess for sale a Florida Cooter, Barbour's Map Turtle, Loggerhead Musk Turtle, Alligator Snapping Turtle, box turtle, or Florida Pine Snake. No permit or license is required to take frogs as long as they are not sold; to take frogs for the purpose of sale, you need a Commercial Fishing License.

# GEORGIA

Department of Natural Resources
Wildlife Resources Division
2070 U.S. Highway 278, S.E.
Social Circle, GA 30279
Phone: (404) 918-6400

**ENDANGERED:**
- Leatherback Sea Turtle *(Dermochelys coriacea)*
- Hawksbill Sea Turtle *(Eretmochelys imbricata)*
- Atlantic Ridley Sea Turtle *(Lepidochelys kempii)*

**THREATENED:**
- Loggerhead Sea Turtle *(Caretta caretta)*
- Green Sea Turtle *(Chelonia mydas)*
- Bog Turtle *(Clemmys muhlenbergii)*
- Gopher Tortoise *(Gopherus polyphemus)*
- Barbour's Map Turtle *(Graptemys barbouri)*
- Alligator Snapping Turtle *(Macroclemys temminckii)*
- Eastern Indigo Snake *(Drymarchon corais couperi)*
- Georgia Blind Salamander *(Haideotriton wallacei)*

**RARE:**
- Map Turtle *(Graptemys geographica)*
- Alabama Map Turtle *(Graptemys pulchra)*
- Flatwoods Salamander *(Ambystoma cingulatum)*
- One-toed Amphiuma *(Amphiuma pholeter)*
- Green Salamander *(Aneides aeneus)*
- Hellbender *(Cryptobranchus alleganiensis)*
- Striped Newt *(Notophthalmus perstriatus)*
- Pigeon Mountain Salamander *(Plethodon petraeus)*

**REGULATIONS:**
In Georgia, certain reptiles and amphibians are unprotected, and may be collected and possessed without any type of permit. This would include native poisonous reptiles, frogs, spring lizards, and freshwater turtles—except for any species listed as Threatened or Endangered, such as the Alligator Snapping Turtle or Flatwoods Salamander. The Copperhead, Cottonmouth, Pigmy Rattlesnake, Timber Rattlesnake, Eastern Diamondback Rattlesnake, and Eastern Coral Snake comprise the list of native, unprotected snakes.

All other species of native reptiles and amphibians are protected under Georgia law and may not be collected or possessed without a permit. A Scientific Collecting Permit must be obtained before any protected species may be taken from the wild. These are only issued for scientific or educational purposes. Protected species may only be possessed after a Wildlife Exhibition Permit has been obtained. Georgia does not regulate the possession of nonvenomous, non-native reptiles such as boa constrictors, pythons, etc. It does, however, regulate all inherently dangerous exotic reptiles, such as cobras, crocodiles and caimans, and Gila Monsters.

# *HAWAII*

Department of Land and Natural Resources
Division of Forestry and Wildlife
1151 Punchbowl St., Room 330
Honolulu, HI 96813
Phone: (808) 587-0166

**ENDANGERED:**
    Pacific Hawksbill Turtle *(Eretmochelys imbricata bissa)*
    Pacific Leatherback Sea Turtle *(Dermochelys coriacea schelegelii)*

**THREATENED:**
    Olive Ridley Sea Turtle *(Lepidochelys olivacea)*
    Pacific Green Sea Turtle *(Chelonia mydas agassizi)*

**REGULATIONS:**
It is illegal to catch, possess, injure, kill, destroy, sell, or offer for sale, transport, or export any indigenous wildlife. It is illegal to possess, process, sell, or offer for sale, transport or export any Endangered or Threatened species. No person shall remove, damage, or disturb the nest of any indigenous species.

Authorized, qualified persons may apply for a Scientific Collecting Permit through the Department of Land and Natural Resouces; contact the Wildlife Biologist.

The only species listed as Endangered or Threatened are sea turtles; Hawaii does not list any amphibians or land reptiles as Endangered or Threatened.

Hawaii has strict regulations prohibiting the importation of specific non-domestic animals that are detrimental to the agriculture and aquaculture industries, natural resources, and environment of the state. All importation of animals shall be by permit. A permit application shall be submitted to the chief with the following information: name and address of shipper and importer; approximate number and kind; sex; object of importation; mode of transportaion; and the approximate date of arrival. Since Hawaii has had problems with alien species, animals prohibited entry are listed below.

Prohibited animals: all *Amphiuma* species, all *Necturus* species, *Siren intermedia*, *Siren lacertina*, all *Phyllobates* species, *Hyla septentrionalis*, all members of the Family Pipidae (except *Pipa pipa* and *Xenopus laevis*, which may be used for research and exhibition by government agencies), all snakes, and *Heloderma* species.

The introduction into Hawaii of live animals is only allowed: (1) for those animals on the list designated as "Conditionally Approved" dated July 15, 1993; (2) by permit approved by the board or chief; (3) after securing appropriate bond for certain animals. The state also lists restricted animals, which are broken down into those species to be used for research and exhibition, and those for private and commercial use. Restricted list animals require a permit for both import and possession. Where a permit for a restricted list animal allows transfer or sale, the proposed transferee must first obtain a permit for possession of the animal by application to the chief, site inspection approval and satisfaction of any bond or other requirements applicable.

# IDAHO

Department of Fish and Game
Nongame & Endangered Wildlife Program
600 S. Walnut St., P.O. Box 25
Boise, ID 83707
Phone: (208) 334-3700

**ENDANGERED:**
Idaho has no herptiles listed as Endangered or Threatened. Their Species of Special Concern are ranked as follows:

**PRIORITY SPECIES:**
Coeur d'Alene Salamander *(Plethodon idahoensis)*

**PERIPHERAL SPECIES:**
Wood Frog *(Rana sylvatica)*
Mojave Black-collared Lizard *(Crotaphytus bicinctores)*
Longnose Snake *(Hypsiglena torquata)*
Western Ground Snake *(Sonora semiannulata)*

**UNDETERMINED STATUS SPECIES:**
Western Pond Turtle *(Clemmys marmorata)*
Northern Alligator Lizard *(Elgaria coerulea)*
Ringneck Snake *(Diadophis punctatus)*
Smooth Green Snake *(Opheodrys vernalis)*

**REGULATIONS:**
No person shall capture alive or hold in captivity at any time more than four Idaho native reptiles or amphibians of any one species except as authorized by Commission regulation or in writing by the Director.

No person shall take or possess those species of wildlife classified as Species of Special Concern at any time or in any manner, except with special permission from the Commission.

For more information on Idaho's herptiles, write to the Nongame & Endangered Wildlife Program for *Idaho's Amphibians & Reptiles: Description, Habitat & Ecology*, Nongame Wildlife Leaflet #7.

# ILLINOIS

Department of Conservation
Lincoln Tower Plaza, 524 S. Second St.
Springfield, IL 62701
Phone: (217) 782-6302

**ENDANGERED:**
- Silvery Salamander *(Ambystoma platineum)*
- Hellbender *(Cryptobranchus alleganiensis)*
- Dusky Salamander *(Desmognathus fuscus)*
- Spotted Turtle *(Clemmys guttata)*
- Illinois Mud Turtle *(Kinosternon flavescens)*
- River Cooter *(Pseudemys concinna)*
- Eastern Massasauga *(Sistrurus catenatus)*
- Eastern Ribbon Snake *(Thamnophis sauritus)*
- Broad-banded Water Snake *(Nerodia fasciata)*

**THREATENED:**
- Illinois Chorus Frog *(Pseudacris streckeri)*
- Four-toed Salamander *(Hemidactylium scutatum)*
- Alligator Snapping Turtle *(Macroclemys temmincki)*
- Kirtland's Snake *(Clonophis kirtlandi)*
- Timber Rattlesnake *(Crotalus horridus)*
- Great Plains Rat Snake *(Elaphe guttata emoryi)*
- Western Hognose Snake *(Heterodon nasicus)*
- Coachwhip Snake *(Masticophis flagellum)*
- Green Water Snake *(Nerodia cyclopion)*

**REGULATIONS:**

It is unlawful to take, possess, buy, sell, offer to buy or sell or barter any reptile, amphibian, or their eggs or parts taken from the wild in Illinois for commercial purposes unless otherwise authorized by statute.

Only those persons who hold a valid sport fishing license may take or attempt to take turtles and/or frogs. Turtles may be taken only by hand, hook and line, or dip net. Bullfrogs may be taken only by hook and line, gig, spear, bow and arrow, hand, or dip net. No person shall take or possess any species of reptile or amphibian listed as Endangered or Threatened in Illinois, except with special permission from the Department of Conservation. All other species of reptiles and amphibians may be captured by any device or method which is not designated or intended to bring about the death or serious injury of the animals captured. This shall not restrict the use of legally taken reptiles or amphibians as bait by anglers. Any captured reptiles or amphibians which are not to be retained in the possession of the captor shall be immediately released at the site of capture.

The daily catch limit for reptiles is eight of each species and for amphibians is eight of each species. The possession limit for reptiles is 16 of each species and for amphibians is 16 of each species. Captive-born offspring from a legally held reptile or amphibian, not intended for commercial purposes, is exempt from the possession limits for a period of 90 days.

# INDIANA

Division of Fish & Wildlife
Nongame & Endgangered Wildlife Program
402 W. Washington St.
Indianapolis, IN 46204-2267
Phone: (317) 232-4080

**ENDANGERED:**
 Hellbender *(Cryptobranchus alleganiensis alleganiensis)*
 Northern Red Salamander *(Pseudotriton ruber ruber)*
 Alligator Snapping Turtle *(Macroclemys temminckii)*
 Hieroglyphic River Cooter *(Pseudemys concinna hieroclyphica)*

**THREATENED:**
 Four-toed Salamander *(Hemidactylium scutatum)*
 Northern Crawfish Frog *(Rana areolata circulosa)*
 Spotted Turtle *(Clemmys guttata)*
 Eastern Mud Turtle *(Kinosternon subrubrum subrubrum)*
 Northern Scarlet Snake *(Cemophora coccinea copei)*
 Kirtland's Snake *(Clonophis kirtlandii)*
 Copperbelly Water Snake *(Nerodia erythrogaster neglecta)*
 Smooth Green Snake *(Opheodrys vernalis)*
 Southeastern Crowned Snake *(Tantilla coronata)*
 Butler's Garter Snake *(Thamnophis butleri)*
 Western Cottonmouth *(Agkistrodon piscivorus leucostoma)*
 Timber Rattlesnake *(Crotalus horridus)*
 Eastern Massasauga *(Sistrurus catenatus catenatus)*

**SPECIES OF SPECIAL CONCERN:**
 Blanding's Turtle
 Ornate Box Turtle
 Rough Green Snake
 Western Ribbon Snake
 Blue-spotted Salamander
 Mudpuppy
 Eastern Spadefoot Toad
 Plains Leopard Frog
 Northern Leopard Frog

**REGULATIONS:**
 Vertebrates classified as Endangered or Threatened in Indiana are protected from "taking" pursuant to the Nongame and Endangered Species Act of 1973 and Fish and Wildlife Administration Rules.
 The Department may issue to a properly accredited individual authorizing that person to collect and possess wild animals in Indiana for scientific purposes only. Scientific collectors must apply for a license, pay a fee of $10, and get the signatures of two relevant scientists as references. A report of the collection by species, number, and location of the collection must be supplied within 15 days after the expiration of the license.

# IOWA

Department of Natural Resources
Division of Fish & Wildlife
Wallace State Office
Des Moines, IA 50319
Phone: (515) 281-5145

**ENDANGERED:**
- Yellow Mud Turtle *(Kinosternon flavescens)*
- Wood Turtle *(Clemmys insculpta)*
- Great Plains Skink *(Eumeces obsoletus)*
- Slender Glass Lizard *(Ophisaurus attenuatus)*
- Yellow-bellied Water Snake *(Nerodia erythrogaster)*
- Western Hognose Snake *(Heterodon nasicus)*
- Speckled Kingsnake *(Lampropeltis getulus)*
- Copperhead *(Agkistrodon contortrix)*
- Prairie Rattlesnake *(Crotalus viridis)*
- Massasauga *(Sistrurus catenatus)*
- Blue-spotted Salamander *(Ambystoma laterale)*
- Central Newt *(Notophthalmus viridescens)*
- Mudpuppy *(Necturus maculosus)*
- Crawfish Frog *(Rana areolata)*

**THREATENED:**
- Stinkpot Turtle *(Sternotherus odoratus)*
- Ornate Box Turtle *(Terrapene ornata)*
- Earth Snake *(Virginia valeriae)*
- Diamondback Water Snake *(Nerodia rhombifera)*

**SPECIES OF SPECIAL CONCERN:**
- Western Worm Snake *(Carphophis amoenus)*

**REGULATIONS:**
All species of amphibians and reptiles are protected in Iowa with the exception of the Common Garter Snake and the Timber Rattlesnake. A person must hold a permit or license from the Iowa Department of Natural Resources to collect and hold all other species.

# KANSAS

Department of Wildlife & Parks
512 S.E. 25th Ave.
Pratt, KS 67124-8174
Phone: (316) 672-5911
Fax: (316) 672-6020

**ENDANGERED:**
    Cave Salamander *(Eurycea lucifuga)*
    Graybelly Salamander *(Eurycea multiplicata griseogaster)*
    Grotto Salamander *(Typhlotriton spelaeus)*

**THREATENED:**
    Central Newt *(Notophthalmus viridescens louisianensis)*
    Dark-sided Salamander *(Eurycea longicauda melanopleura)*
    Eastern Narrowmouth Toad *(Gastrophryne carolinensis)*
    Green Frog *(Rana clamitans melanota)*
    Northern Spring Peeper *(Pseudacris crucifer crucifer)*
    Strecker's Chorus Frog *(Pseudacris streckeri streckeri)*
    Western Green Toad *(Bufo debilis insidior)*
    Broadhead Skink *(Eumeces laticeps)*
    Checkered Garter Snake *(Thamnophis marcianus marcianus)*
    Common Map Turtle *(Graptemys geographical)*
    New Mexico Blind Snake *(Leptotyphlops dulcis dissectus)*
    Northern Redbelly Snake *(Storeria occipitomaculata occipitumaculata)*
    Texas Longnose Snake *(Rhinocheilus lecontei tessellatus)*
    Texas Night Snake *(Hypsiglena torquata jani)*
    Western Earth Snake *(Virginia valeriae elegans)*

**REGULATIONS:**
    A hunting license is required to take any wildlife species, except invertebrates. No more than five of any one species of amphibian or reptile may be possessed other than for use as fishing bait. The Bullfrog open season is July 1 through October 31, and Common Snapping Turtles and softshelled turtles season is January 1 through December 31. The bag limit for Common Snapping Turtles and softshelled turtles is eight of any combination; possession limit is 24. The daily Bullfrog limit is eight; the possession limit is 24.

    Exotic wildlife (species which are non-migratory and are not native or indigenous to Kansas) may be possessed without limit in time and number. Exotic wildlife must be confined or controlled at all times and shall not be released onto the lands or into the waters of this state.

# KENTUCKY

Department of Fish & Wildlife Resources
Nongame Program
#1 Game Farm Rd.
Frankfort, KY 40601
Phone: (502) 564-5448

**ENDANGERED:**
    Three-toed Amphiuma *(Amphiuma tridactylum)*
    Wehrle's Salamander *(Plethodon wehrlei)*
    Kirtland's Snake *(Clonophis kirtlandii)*
    Broad-banded Water Snake *(Nerodia fasciata confluens)*
    Mississippi Green Water Snake *(Nerodia cyclopion)*
    Southern Coal Skink *(Eumeces anthracinus pluvialis)*

**THREATENED:**
    Three-lined Salamander *(Eurycea guttolineata)*
    Bird-voiced Treefrog *(Hyla avivoca)*
    Northern Coal Skink *(Eumeces anthracinus anthracinus)*
    Eastern Slender Glass Lizard *(Ophisaurus attenuatus longicaugus)*
    Northern Pine Snake *(Pituophis melanoleucus melanoleucus)*
    Western Pigmy Rattlesnake *(Sistrurus miliarius streckeri)*
    Western Ribbon Snake *(Thamnophis proximus proximus)*
    Alligator Snapping Turtle *(Macroclemys temminckii)*

**SPECIES OF SPECIAL CONCERN:**

| | |
|---|---|
| Green Treefrog | Southern Painted Turtle |
| Barking Treefrog | Corn Snake |
| Gray Treefrog | Western Mud Snake |
| Redback Salamander | Scarlet Kingsnake |
| Northern Crawfish Frog | Copperbelly Water Snake |
| Northern Leopard Frog | Eastern Ribbon Snake |
| Midland Smooth Softshell | Southeastern Five-lined Skink |

**REGULATIONS:**
    Kentucky has 105 amphibians and reptiles, none of which are Federally listed as Endangered or Threatened. Kentucky has no official Endangered/Threatened list, and the above list was *suggested* by the Kentucky State Nature Preserves Commission [phone (502) 564-2886]. In order to collect any Endangered or Threatened species in Kentucky, the individuals must be named an "agent of the state." In order to collect herps not Endangered or Threatened, an individual must have either an educational or scientific permit.
    Holders of sport fishing licenses may seine live bait from public waters and may possess up to 100 salamanders, 100 frogs (other than Bullfrogs), and 100 tadpoles. The Bullfrog possession limit is 30. Turtles also may be taken with a fishing license. Unprotected species (includes lizards and snakes) may be taken year-round except November 1 to November 22. Is is illegal to buy, sell, possess, propagate, exhibit, import, or transport any wildlife (includes reptiles) without a permit.

# LOUISIANA

Department of Wildlife & Fisheries
Louisiana Natural Heritage Program
P.O. Box 98000
Baton Rouge, LA 70898-9000
Phone: (504) 765-2800

**ENDANGERED:**
    Hawksbill Sea Turtle *(Eretmochelys imbricata)*
    Kemp's Ridley Sea Turtle *(Lepidochelys kempii)*
    Leatherback Sea Turtle *(Dermochelys coriacea)*

**THREATENED:**
    Green Sea Turtle *(Chelonia mydas)*
    Loggerhead Sea Turtle *(Caretta caretta)*
    Gopher Tortoise *(Gopherus polyphemus)*
    Ringed Sawback Turtle *(Graptemys oculifera)*

**SPECIES OF SPECIAL CONCERN:**

| | |
|---|---|
| Eastern Tiger Salamander | Loggerhead Sea Turtle |
| Two-toed Amphiuma | Green Sea Turtle |
| Southern Two-lined Salamander | Hawksbill Sea Turtle |
| Four-toed Salamander | Kemp's Ridley Sea Turtle |
| Southern Redback Salamander | Alligator Snapping Turtle |
| Webster's Salamander | Leatherback Sea Turtle |
| Louisiana Slimy Salamander | Common Map Turtle |
| Gulf Coast Mud Salamander | Ringed Map Turtle |
| Southern Red Salamander | Alabama Map Turtle |
| Ornate Chorus Frog | Mississippi Diamondback Terrapin |
| Strecker's Chorus Frog | Ornate Box Turtle |
| Dusky Crawfish Frog | Stripe-necked Musk Turtle |
| Eastern Glass Lizard | Gopher Tortoise |
| Southern Prairie Skink | Gulf Coast Smooth Softshell |
| Western Worm Snake | Tan Racer |
| Rainbow Snake | Mole Kingsnake |

**REGULATIONS:**
    All persons engaged in the collection of native reptiles and amphibians for noncommercial purposes must possess a fishing license. All persons engaged in the sale of native reptiles and amphibians collected in Louisiana must possess a collector's license (except those who are 16 years old and younger). Persons engaged in legitimate herpetological research may request a scientific collector's permit.
    All persons engaged in the buying, acquiring, or the handling by any means any species of native reptile or amphibian in Louisiana for resale, or any person engaged in the shipping or transporting of any native reptile or amphibian either into or out of the state must possess a reptile and amphibian wholesale/retail dealer's license.

# MAINE

Department of Inland Fisheries & Wildlife
Wildlife Division
284 State St.
Augusta, ME 04333
Phone: (207) 287-2766

**ENDANGERED:**
  Leatherback Sea Turtle *(Dermochelys coricea)*
  Atlantic Ridley Sea Turtle *(Lepidochelys kempii)*
  Box Turtle *(Terrapene carolina)*
  Black Racer *(Coluber constrictor)*

**THREATENED:**
  Loggerhead Sea Turtle *(Caretta caretta)*
  Blanding's Turtle *(Emydoidea blandingii)*
  Spotted Turtle *(Clemmys guttata)*

**SPECIES OF SPECIAL CONCERN:**
  Ribbon Snake *(Thamnophis sauritus)*

**SPECIES OF INDETERMINATE STATUS:**
  Tremblay's Salamander *(Ambystoma tremblayi)*
  Wood Turtle *(Clemmys insculpta)*
  Brown Snake *(Storeria dekayi)*

**REGULATIONS:**
  Maine has no regulations concerning private (non-commercial) collection of native non-Endangered reptiles and amphibians. A new law was enacted in 1993 prohibiting commercial trade in the state's turtles and snakes. Maine does issue permits to collect Snapping Turtles.
  A Scientific Collection Permit is available to qualified individuals to collect Endangered species for study or to educate the public.

# MARYLAND

Department of Natural Resources
Tawes State Office Building
Annapolis, MD 21401
Phone: (410) 974-2870

**ENDANGERED:**
Eastern Tiger Salamander *(Ambystoma tigrinum)*
Green Salamander *(Aneides aeneus)*
Hellbender *(Cryptobranchus alleganiensis)*
Eastern Narrow-mouthed Toad *(Gastrophryne carolinensis)*
Barking Treefrog *(Hyla gratiosa)*
Atlantic Leatherback Turtle *(Dermochelys coriacea)*
Atlantic Hawksbill Sea Turtle *(Eretmochelys imbricata)*
Atlantic Ridley Sea Turtle *(Lepidochelys kempii)*
Mountain Earth Snake *(Virginia valeriae pulchra)*
Northern Coal Skink *(Eumeces anthracinus)*

**THREATENED:**
Atlantic Loggerhead Sea Turtle *(Caretta caretta)*
Atlantic Green Sea Turtle *(Chelonia mydas)*

**IN NEED OF CONSERVATION:**
Wehrle's Salamander *(Plethodon wehrlei)*
Mountain Chorus Frog *(Pseudacris brachyphona)*
Carpenter Frog *(Rana virgatipes)*
Eastern Spiny Softshell Turtle *(Apalone spinifera)*
Map Turtle *(Graptemys geographica)*

**REGULATIONS:**
Without a permit, a person: may not possess more than four of each individual native Maryland reptile or salamander, live or dead; may not possess more than four adults and 25 eggs or tadpoles of each individual native frog or toad, live or dead; may possess only one of each individual native turtle, live or dead, and only Eastern Box Turtles may have been obtained from the wild. A person without a permit (but with a valid fishing license) may possess not more than 25 individual amphibians in total for use as bait.

A permit is necessary in order to breed, attempt to breed, sell, offer for sale, trade, or barter any native reptile or amphibian. Permittees may possess an unlimited number of any native reptiles or amphibians that are captive-produced or legally obtained from outside of Maryland. Copperheads may be held under the authority of a permit issued. Turtles must have a carapace length of at least four inches. A permittee may not breed turtles.

A reptile or amphibian that has been captively produced or that is not native to Maryland may not be released into the wild.

# MASSACHUSETTS

Division of Fisheries & Wildlife
100 Cambridge St.
Boston, MA 02202
Phone: (617) 727-9194

**ENDANGERED:**
    Bog Turtle *(Clemmys muhlenbergii)*
    Plymouth Redbelly Turtle *(Pseudemys rubriventris bangsi)*
    Hawksbill Sea Turtle *(Eretmochelys imbricata)*
    Atlantic Ridley Sea Turtle *(Lepidochelys kempii)*
    Leatherback Sea Turtle *(Dermochelys coriacea)*
    Black Rat Snake *(Elaphe obsoleta)*
    Copperhead *(Agkistrodon contortrix)*
    Timber Rattlesnake *(Crotalus horridus)*

**THREATENED:**
    Marbled Salamander *(Ambystoma opacum)*
    Eastern Spadefoot Toad *(Scaphiopus holbrookii)*
    Blandings Turtle *(Emydoidea blandingii)*
    Diamondback Terrapin *(Malaclemys terrapin)*
    Loggerhead Sea Turtle *(Caretta caretta)*
    Green Sea Turtle *(Chelonia mydas)*
    Worm Snake *(Carphophis amoenus)*

**SPECIES OF SPECIAL CONCERN:**

| | |
|---|---|
| Jefferson Salamander | Spotted Turtle |
| Blue-spotted Salamander | Wood Turtle |
| Spring Salamander | Eastern Box Turtle |
| Four-toed Salamander | |

**REGULATIONS:**
    Reptiles and amphibians (except Bullfrogs, Green Frogs, and Snapping Turtles) may be hunted, fished, trapped, or taken from January 1 through December 31, up to a possession limit of two, except that no reptile or amphibian may be taken by hunting with firearms or bow and arrow on any Sunday. Bullfrogs and Green Frogs (except eggs) larger than three inches may be hunted or taken from July 16 to September 30. Not more than 12 frogs (singly or in aggregate) of either species can be taken, or more than 24 in possession. There is no daily or seasonal bag limit on Snapping Turtles; season is January 1 through December 31. Unless authorized in a permit, no herp may be taken from the wild for purposes of sale; Snapping Turtles taken under such a permit must be greater than six inches in carapace length. The Endangered, Threatened, and species of special concern listed above shall not be disturbed or harrassed, hunted, fished, trapped, or taken. A Class 4 Propagator's License authorizes a person to possess, maintain, and propagate for purposes of sale or barter specified reptiles or amphibians. A Class 7 Possessor's License authorizes a person to possess reptiles or amphibians.

# MICHIGAN

Department of Natural Resources
Wildlife Division, Nongame Program
Box 30180
Lansing, MI 48909
Phone: (517) 373-1263

**ENDANGERED:**
    Smallmouth Salamander *(Ambystoma texanum)*
    Kirtland's Snake *(Clonophis kirtlandii)*
    Copperbelly Water Snake *(Nerodia erythrogaster neglecta)*

**THREATENED:**
    Marbled Salamander *(Ambystoma opacum)*
    Eastern Fox Snake *(Elaphe vulpina gloydi)*

**SPECIES OF SPECIAL CONCERN:**
    Blanchard's Cricket Frog *(Acris crepitans blanchardi)*
    Boreal Chorus Frog *(Pseudacris triseriata maculata)*
    Spotted Turtle *(Clemmys guttata)*
    Wood Turtle *(Clemmys insculpta)*
    Eastern Box Turtle *(Terrapene carolina)*
    Black Rat Snake *(Elaphe obsoleta)*

**REGULATIONS:**
    Michigan's frogs and turtles are covered under the state's fish regulations. Open season for frogs is the Saturday before Memorial Day to November 15, with no size or limit restrictions. Frogs may be taken and/or sold without a license. Turtles have no closed season, and there are no size or limit restrictions.
    Before 1988, the Michigan D.N.R. only had jurisdiction over frogs; since the 1988 change, the D.N.R. now has jurisdiction over all herptiles. The Department now totally protects the Black Rat Snake, Cricket Frog (becoming rare in the state), Wood Turtle, Eastern Box Turtle, Spotted Turtle, and Blanding's Turtle.
    Any turtles other than Snapping Turtles and the protected species listed above have a daily limit of 2 and possession limit of 6; there is no closed season. Snapping Turtles have a daily limit of 6 for personal use, or 10 with a commercial license. Snapping Turtle open season varies with the region: Upper Peninsula—July 15 to September 30; Northern Lower Peninsula—July 1 to September 30; Southern Lower Peninsula—June 23 to September 30. Amphibian possession limit is 12 of each species.

# MINNESOTA

Department of Natural Resources
Nongame Wildlife Program
Box 7, 500 Lafayette Rd.
St. Paul, MN 55155-4001
Phone: (612) 296-6157

**ENDANGERED:**
    Five-lined Skink *(Eumeces fasciatus)*

**THREATENED:**
    Wood Turtle *(Clemmys insculpta)*
    Blanding's Turtle *(Emydoidea blandingi)*

**SPECIES OF SPECIAL CONCERN:**
| | |
|---|---|
| Snapping Turtle | Milk Snake |
| Blue Racer | Gopher Snake |
| Timber Rattlesnake | Massasauga |
| Black Rat Snake | Lined Snake |
| Fox Snake | Northern Cricket Frog |
| Western Hognose Snake | Bullfrog |
| Eastern Hognose Snake | Pickerel Frog |

**REGULATIONS:**
    Endangered or Threatened species are legally protected. Permits issued by the DNR are required to legally take, possess, import, transport, purchase, sell, or dispose of these species; permits may be issued for scientific and educational purposes and rehabilitation. Permits are not required for taking/possessing species of special concern, and all other lizards, snakes, salamanders, and toads not officially listed in Minnesota are totally unprotected. Minnesota does have regulations for turtles and frogs:

    Turtles—*Noncommercial*—Any person permitted by law to take fish by angling may take, possess, buy, sell, and transport turtles. Turtles may not be taken by the use of explosives, drugs, poisons, lime, or other deleterious substances or by the use of nets (other than landing nets or traps). Possession limit for snapping turtles is 10 and the dorsal surface of the carapace must be 10 inches or more across at its narrowest point. *Commercial*—A $50 Commercial Turtle License is necessary to take, transport, purchase, and possess for sale unprocessed turtles within the state, without limit. Turtle traps, turtle hooks, and commercial fishing nets may be used by holders of Commercial Turtle Licenses and are the only gear to be used.

    Frogs—Any person permitted by law to take fish by angling may take or possess frogs for bait purposes only. Frogs may not be taken for bait if they exceed six inches from the tip of the nose to the tip of the hind legs (when the hind legs are fully extended). Bait frogs can be possessed, bought, sold, and transported in any numbers. Frogs may not be taken from April 1 to May 15. No more than 150 frogs over six inches in length may be possessed in or transported through the state except by common carrier. The taking, possessing, purchasing, transporting, or selling of frogs for purposes other than as bait within the state is prohibited. Scientific or special permits may be issued to educational and scientific institutions within Minnesota.

# MISSISSIPPI

Department of Wildlife, Fisheries and Parks
Museum of Natural History
111 N. Jefferson St.
Jackson, MS 39202
Phone: (601) 354-7303

**ENDANGERED:**
    Dusky Gopher Frog *(Rana capito sevosa)*
    Cave Salamander *(Eurycea lucifuga)*
    Green Salamander *(Aneides aeneus)*
    Spring Salamander *(Gyrinophilus porphyriticus)*
    Black Pine Snake *(Pituophis melanoleucus lodingi)*
    Eastern Indigo Snake *(Drymarchon corais couperi)*
    Rainbow Snake *(Farancia erytrogramma)*
    Southern Hognose Snake *(Heterodon simus)*
    Atlantic Ridley Sea Turtle *(Leopidochelys kempii)*
    Black-knobbed Sawback Turtle *(Graptemys nigrinoda)*
    Gopher Tortoise *(Gopherus polyphemus)*
    Green Sea Turtle *(Chelonia mydas)*
    Hawksbill Sea Turtle *(Eretmochelys imbricata)*
    Leatherback Sea Turtle *(Dermochelys coriacea)*
    Loggerhead Sea Turtle *(Caretta caretta)*
    Mississippi Redbelly Turtle *(Pseudemys* sp.)
    Ringed Sawback Turtle *(Graptemys oculifera)*
    Yellow-blotched Sawback *(Graptemys flavimaculata)*

**REGULATIONS:**
    Endangered species may not be possessed without special permits from the Department. Species deemed in need of management (which includes almost all other species of reptiles and amphibians) may not enter into commercial trade unless they have been propagated in captivity by an individual holding a captive propagation permit. Species deemed in need of management may be possessed for personal use by an individual with the appropriate hunting license. For state residents, cost of the license is approximately $16. For nonresidents, costs will vary according to the state in which the person resides, but usually average approximately $125. The bag limits for personal use are four specimens of any species or subspecies of reptile, not to exceed more than a total of 20 reptiles. Limits on amphibians are 40 specimens, with no more than four of any single species or subspecies. Mississippi does not regulate exotic species of herps at the state level, but many cities have local ordinances concerning "dangerous" wildlife held in captivity. Bullfrogs (including *Rana catesbeiana, R. grylio,* and *R. clamintans*) are considered game species and have a specific season and bag limits which are set on a yearly basis. The Common Snapper *(Chelydra serpentina)* is the only species of reptile or amphibian in Mississippi which is considered a commercial species. Licenses to take Common Snappers cost approximately $200.

# MISSOURI

Department of Conservation
2901 W. Truman Blvd.
P.O. Box 180
Jefferson City, MO 65102-0180
Phone: (314) 751-4115

**ENDANGERED:**
    Western Chicken Turtle *(Deirochelys reticularia miaria)*
    Blanding's Turtle *(Emydoidea blandingii)*
    Yellow Mud Turtle *(Kinosternon flavescens flavescens)*
    Illinois Mud Turtle *(Kinosternon flavescens spooneri)*
    Western Fox Snake *(Elaphe vulpina vulpina)*
    Western Smooth Green Snake *(Opheodrys vernalis blanchardi)*
    Eastern Massasauga *(Sistrurus catenatus catenatus)*
    Western Massasauga *(Sistrurus catenatus tergeminus)*

**RARE:**
    Mole Salamander *(Ambystoma talpoideum)*
    Four-toed Salamander *(Hemidactylium scutatum)*
    Illinois Chorus Frog *(Pseudacris streckeri illinoensis)*
    Northern Leopard Frog *(Rana pipiens)*
    Wood Frog *(Rana sylvatica)*
    Eastern Spadefoot Toad *(Scaphiopus holbrookii holbrookii)*
    Alligator Snapping Turtle *(Macroclemys temminckii)*
    Northern Scarlet Snake *(Cemophora coccinea copei)*
    Dusty Hognose Snake *(Heterodon nasicus gloydi)*
    Plains Hognose Snake *(Heterodon nasicus nasicus)*
    Texas Horned Lizard *(Phrynosoma cornutum)*
    Great Plains Skink *(Eumeces obsoletus)*

**REGULATIONS:**
    Nearly all 108 species of Missouri's amphibians and reptiles are considered nongame—there is no open season on them. Game species are: Bullfrogs, Green Frogs, Common Snapping Turtles, and both softshell species. Also, frogs, other than rare or endangered species, can be taken for fish bait with a valid fishing permit. There is no provision in the state's regulation allowing the taking of amphibians and reptiles (except Common Snapping Turtles and softshells from commercial fishing waters) for commercial sale. Thus, the majority of Missouri's amphibians and reptiles cannot be sold or given away. A resident of Missouri is allowed to take up to five live, nongame specimens for personal use as a pet or for captive study; however, these specimens cannot be sold or given away. A Missourian who intends to captive-breed ANY amphibian or reptile species native to Missouri must apply for a Class I Wildlife Breeder Permit. Also, their breeding stock must be documented to come from a population outside of Missouri. The annual cost of this permit is $50. As of January 1, 1994, a Missourian keeping venomous reptiles OF ANY KIND alive in captivity must apply for a CLASS II Wildlife Breeder Permit (which covers dangerous animals). The annual cost of this permit is $150.

# MONTANA

Department of Fish, Wildlife & Parks
Research and Technical Services Bureau
Montana State University Campus
Bozeman, MT 59717-0322
Phone: (406) 444-2535

**ENDANGERED:**
There are no Endangered or Threatened species of reptiles or amphibians listed in Montana at this time.

**SPECIES OF SPECIAL INTEREST AND CONCERN:**
Snapping Turtle *(Chelydra serpentina)*
Spiny Softshell Turtle *(Trionyx spiniferus)*
Smooth Green Snake *(Opheodrys vernalis)*
Western Hognose Snake *(Heterodon nasicus)*
Cour d'Alene Salamander *(Plethodon idahoensis)*
Pacific Giant Salamander *(Dicamptodon ensatus)*
Tailed Frog *(Ascaphus truei)*
Canadian Toad *(Bufo hemiophrys)*
Wood Frog *(Rana sylvatica)*

**REGULATIONS:**
Montana does not have any rules, regulations, or laws which deal specifically with protection of reptiles or amphibians. However, state law does provide the option of establishing regulations by designating nongame species "in need of management." It shall be unlawful for any person to take, possess, transport, export, sell, or offer for sale nongame wildlife deemed by the Department to be in need of management.

The Department conducts ongoing investigations of nongame wildlife to update their status.

# NEBRASKA

Game and Parks Commission
2200 N. 33rd St.
P.O. Box 30370
Lincoln, NE 68503-0370
Phone: (402) 471-0641

**ENDANGERED:**
Nebraska has no amphibians or reptiles currently on their Endangered or Threatened species lists.

**SPECIES IN NEED OF CONSERVATION:**
Short-horned Lizard *(Phrynosoma douglassii)*

**REGULATIONS:**
The laws and regulations pertaining to amphibians and reptiles in Nebraska are somewhat unclear at the present time because a new statute just became law on January 1, 1994, and the agency has not yet adopted supporting regulations. To summarize the situation before January 1, 1994:

Several different sections of Nebraska Law and Game and Parks Commission Regulations apply to the approximately 60 species of herps found in Nebraska. The Short-horned Lizard is listed as a Nongame Species in Need of Conservation and it is unlawful to take or possess members of that species. The Snapping Turtle, Bullfrog, and Tiger Salamander are considered game species and a fishing permit is required to take those species. Regulations, including seasons, bag limits, and methods of take for those three species are listed in the current Nebraska Fishing Guide. A bait vendor's permit is required for both residents and nonresidents over 16 years of age who take salamanders and Leopard or Striped Frogs to sell as bait for a profit. Except for the species listed above, all other amphibians and reptiles are considered nongame species. They are not protected by seasons, bag limits, or specified methods of take. Residents are allowed to take or possess these species without any type of permit; however, nonresidents (regardless of age) are required to possess a small-game hunting permit to take or possess any amphibian or turtle.

The new statute will primarily affect the commercial exploitation of amphibians and reptiles and the export from the state. The Legislature wants to prevent the importation into Nebraska of any live reptile or amphibian which may cause economic or ecologic harm or be injurious to human beings, agriculture, horticulture, forestry, or wildlife of the state; they also want to prevent the commercial exploitation of any dead or live reptile or amphibian taken from the wild. It will be unlawful for any person to import into the state or release to the wild any live reptile or amphibian (including eggs) except those which are approved by rules and regulations of the commission. Special considerations will be given to zoos, parks, exhibits, and other such institutions for educational and scientific purposes.

# NEVADA

Department of Wildlife
1100 Valley Rd.
P.O. Box 10678
Reno, NV 89520-0022
Phone: (702) 688-1500

**PROTECTED:**
    Gila Monster *(Heloderma suspectum)*

**THREATENED:**
    Desert Tortoise *(Gopherus agassizi)*

**SPECIES OF SPECIAL CONCERN:**

| | |
|---|---|
| Chuckwalla | Rubber Boa |
| Short-horned Lizard | Rosy Boa |
| Tree Lizard | Sonoran Mountain Kingsnake |
| Long-tailed Brush Lizard | Ringneck Snake |
| Gilbert Skink | Worm Snake |
| Western Skink | Lyre Snake |
| Fringe-toed Lizard | Western Diamondback Rattlesnake |
| Desert Iguana | Black-tailed Rattlesnake |
| Banded Gecko | Amargosa Toad |

**REGULATIONS:**
    The importation, transportation, or possession of the following species of live wildlife or hybrids thereof, including viable embryos or gametes, is prohibited: alligators and caimans; crocodiles; the Gharial; all bird snakes (genus *Thelotornis*); the Boomslang; all keelbacks (genus *Rhabdophis*); burrowing asps (Family Atractaspidae); all species in the Family Elapidae, except species in the subfamily Hydrophiinae; all species in the Family Viperidae, except species indigenous to Nevada; snapping turtles; clawed frogs (genus *Xenopus*); giant or marine toads *(Bufo horribilis, B. marinus, B. paracnemis)*.
    It is unlawful, except by the written consent and approval of the Division, for any person to receive or remove from one stream or body of water in this state to any other, or from one portion of the state to any other, or to any other state, any aquatic life, wildlife, spawn, eggs, or young of any of them. The Division shall require an applicant to conduct an investigation to confirm that such an introduction or removal will not be detrimental to the wildlife or the habitat of wildlife in this state. The Bullfrog is classified as a Game Species.
    No person may: possess any live wildlife unless licensed by the Division to do so; capture live wildlife in this state to stock a commercial or non-commercial wildlife facility. The Division may issue commercial and non-commercial licenses for the possession of live wildlife upon receipt of the applicable fee.

# NEW HAMPSHIRE

Fish and Game Department
2 Hazen Dr.
Concord, NH 03301
Phone: (603) 271-3421

**ENDANGERED:**
    Timber Rattlesnake *(Crotalus horridus)*

**SPECIES OF SPECIAL CONCERN:**
    Spotted Turtle *(Clemmys guttata)*
    Blanding's Turtle *(Emydoidea blandingii)*
    Wood Turtle *(Clemmys insculpta)*

**REGULATIONS:**
    Any amphibian and any reptile which is non-venomous is listed as non-controlled, which means that the Department does not regulate any activities pertaining to them. As far as venomous reptiles are concerned, the only activity that would be allowed under Department rules is the possession of the animals for the purpose of *exhibition only*, which requires an exhibition permit from this Department. Venomous reptiles and amphibians cannot be propagated.

    Although New Hampshire currently does not regulate amphibian and non-venomous reptile species, the state is currently assessing strategies necessary to protect certain herptile populations within the state. Several research projects have been conducted to determine the status and distribution of Spotted Turtles, Blanding's Turtles, and Wood Turtles.

    No wildlife shall be propagated unless it has been designated as controlled and a propagator has been permitted to possess it for the activity of propagation and sale of the designated species.

    Permittee categories: (1) an individual person; (2) a propagator; (3) a non-profit educational organization and wildlife educational institute; (4) a person operating a hunting preserve; (5) an exhibitor. No permit is required to possess or import non-controlled species.

# NEW JERSEY

Division of Fish, Game and Wildlife
CN 400
Trenton, NJ 08625-0400
Phone: (609) 292-2965

**ENDANGERED:**
- Tremblay's Salamander (*Ambystoma tremblayi*)
- Blue-spotted Salamander (*Ambystoma laterale*)
- Eastern Tiger Salamander (*Ambystoma tigrinum tigrinum*)
- Pine Barrens Treefrog (*Hyla andersonii*)
- Southern Gray Treefrog (*Hyla chrysocelis*)
- Bog Turtle (*Clemmys muhlenbergi*)
- Atlantic Hawksbill Sea Turtle (*Eretmochelys imbricata*)
- Atlantic Loggerhead Sea Turtle (*Caretta caretta*)
- Atlantic Ridley Sea Turtle (*Lepidochelys kempii*)
- Atlantic Leatherback Sea Turtle (*Dermochelys coriacea*)
- Corn Snake (*Elaphe guttata guttata*)
- Timber Rattlesnake (*Crotalus horridus horridus*)

**THREATENED:**
- Long-tailed Salamander (*Eurycea longicauda*)
- Eastern Mud Salamander (*Pseudotriton montanus*)
- Wood Turtle (*Clemmys insculpta*)
- Atlantic Green Sea Turtle (*Chelonia mydas*)
- Northern Pine Snake (*Pituophis melanoleucus melanoleucus*)

**REGULATIONS:**
No person shall have in possession any nongame species or exotic species of any reptile or amphibian without a permit. Exotic species and nongame species requiring a permit for possession include BUT ARE NOT LIMITED TO: pythons, rat snakes, boas, kingsnakes, Racers, Ringneck Snakes, green snakes, the Collared Lizard, monitors, skinks, ameivas, chuckwallas, alligator lizards, geckos, and the Armadillo Lizard. The following species may be possessed without a permit: the American Anole, Common Green Iguana, Boa Constrictor, Eastern Painted Turtle, Snapping Turtle, Fence Lizards (*Sceloporus occidentalis* and *undulatus*), garter snakes, Tokay Gecko, Ribbon Snake (except *T. sirtalis tetrataenia*), Leopard Frog, Green Frog, American Toad, Fowler's Toad, Bullfrog, Red-spotted Newt, and the Dusky Salamander. Potentially dangerous species require a special permit as well, including: Helodermatidae, Elapidae, Viperidae, Crotalidae, alligators, caimans, crocodiles, and the Gavial.

The albino form of the Corn Snake, or Red Rat Snake, was recently allowed by regulation to be sold or possessed in New Jersey. Keep in mind, however, that no other form of this Endangered species can be used commercially, or possessed for non-commercial reasons without the proper permit, or transferred to another individual in this state without authorization. Also, breeding of albino Red Rat Snakes is not recommended by New Jersey due to the possibility of producing individuals with a normal-colored appearance. These would not be legal to possess and problems could result. Only clearly and completely albino specimens of this species are allowed for sale and possession.

# NEW MEXICO

Department of Game and Fish
Villagra Building
P.O. Box 25112
Santa Fe, NM 87504
Phone: (505) 827-7911

**ENDANGERED GROUP 1:**
*(species whose prospects of survival in New Mexico are in jeopardy)*
- Gila Monster *(Heloderma suspectum)*
- Bunch Grass Lizard *(Sceloporus scalaris)*
- Gray-checkered Whiptail *(Cnemidophorus dixoni)*
- Giant Spotted Whiptail *(Cnemidophorus burti)*
- Mexican Garter Snake *(Thamnophis eques)*
- Ridgenose Rattlesnake *(Crotalus willardi)*
- Lowland Leopard Frog *(Rana yavapaiensis)*
- Western Toad *(Bufo boreas)*
- Great Plains Narrowmouth Toad *(Gastrophryne olivacea)*
- Spotted Chorus Frog *(Pseudacris clarkii)*

**ENDANGERED GROUP 2:**
*(species whose prospects of survival in New Mexico are likely to be in jeopardy within the forseeable future)*
- River Cooter *(Pseudemys concinna)*
- Dunes Sagebrush Lizard *(Sceloporus graciosus arenicolous)*
- Mountain Skink *(Eumeces callicephalus)*
- Plainbelly Water Snake *(Nerodia erythrogaster)*
- Green Rat Snake *(Senticolis triaspis)*
- Narrowhead Garter Snake *(Thamnophis rufipunctatus)*
- Western Ribbon Snake *(Thamnophis proximus)*
- Mottled Rock Rattlesnake *(Crotalus lepidus lepidus)*
- Jemez Mountain Salamander *(Plethodon neomexicanus)*
- Sacramento Mountain Salamander *(Aneides hardii)*
- Colorado River Toad *(Bufo alvarius)*

**REGULATIONS:**
New Mexico law provides for the taking of protected wildlife for scientific and educational purposes. Such purposes are defined as activities that expand scientific knowledge, educate, or similarly promote the conservation of New Mexico's wildlife. Although these activities are among the highest and best uses that humans make of wildlife, they must meet stringent standards to be authorized and permitted by New Mexico Department of Game and Fish. In particular, they must (a) not be harmful to wildlife populations or taxa; (b) be properly justified, conducted, and consummated; (c) be appropriately administered; and (d) not abridge the legitimate rights of others that use wildlife in New Mexico.

# NEW YORK

Department of Environmental Conservation
Division of Fish and Wildlife
50 Wolf Rd.
Albany, NY 12233
Phone: (518) 457-0689

**ENDANGERED:**
- Tiger Salamander *(Ambystoma tigrunum)*
- Bog Turtle *(Clemmys muhlengergi)*
- Leatherback Sea Turtle *(Dermochelys coriacea)*
- Hawksbill Sea Turtle *(Eretmochelys imbricata)*
- Atlantic Ridley Sea Turtle *(Lepidochelys kempii)*
- Massasauga *(Sistrurus catenatus)*

**THREATENED:**
- Cricket Frog *(Acris crepitans)*
- Mud Turtle *(Kinosternon subrubrum)*
- Blanding's Turtle *(Emydoidea blandingi)*
- Loggerhead Sea Turtle *(Caretta caretta)*
- Green Sea Turtle *(Chelonia mydas)*
- Timber Rattlesnake *(Crotalus horridus)*

**SPECIES OF SPECIAL CONCERN:**

| | |
|---|---|
| Southern Leopard Frog | Spotted Turtle |
| Hellbender | Wood Turtle |
| Jefferson Salamander | Diamondback Terrapin |
| Blue-spotted Salamander | Worm Snake |
| Spotted Salamander | Eastern Hognose Snake |

**REGULATIONS:**

Endangered or Threatened species may not be collected, possessed or sold without a license. Special concern species *do not* receive similar protection. Box turtles *(Terrapene* ssp.) and Wood Turtles *(Clemmys insculpta)* receive protection as Game Species with no open season. Neither species may be collected or possessed without a license. Any person who has a small game hunting license, a fishing license, or is entitled to fish without a license may take frogs. A small game license is required to take frogs with a gun. Frogs may be taken in any number from June 16-September 30 between sunrise and sunset.

No person shall liberate any species except under permit. No reptiles may be collected in the state without a license. Licenses will only be issued for propagation, scientific, or exhibition purposes. Unprotected species may be possessed without a license provided they were legally obtained. The Department's policy is not to issue a Scientific Collector's License to individuals who wish to keep protected species as pets. Public Health Law regulates the sale of unprotected turtles in New York; turtles with a carapace length of greater than four inches may be sold. Diamondback Terrapin may be taken from August 1-April 30 with a Diamondback Terrapin License. Diamondbacks with a carapace length less than four inches or greater than seven inches are not allowed to be taken.

# NORTH CAROLINA

Division of Wildlife Management
Archdale Building
Raleigh, NC 27604
Phone: (919) 733-3391

**ENDANGERED:**
    Hawksbill Sea Turtle *(Eretmochelys imbricata)*
    Atlantic Ridley Sea Turtle *(Lepidochelys kempii)*
    Leatherback Sea Turtle *(Dermochelys coriacea)*
    Green Salamander *(Aneides aeneus)*

**THREATENED:**
    American Alligator *(Alligator mississippiensis)*
    Bog Turtle *(Clemmys muhlenbergii)*
    Green Sea Turtle *(Chelonia mydas)*
    Loggerhead Sea Turtle *(Caretta caretta)*
    Eastern Tiger Salamander *(Ambystoma tigrinum)*
    Wehrle's Salamander *(Plethodon wehrlei)*

**SPECIES OF SPECIAL CONCERN:**

| | |
|---|---|
| Carolina Salt Marsh Snake | Crevice Salamander |
| Eastern Smooth Green Snake | Dwarf Salamander [silver] |
| Northern Pine Snake | Eastern Hellbender |
| Outer Banks Kingsnake | Four-toed Salamander |
| Eastern Spiny Softshell Turtle | Junaluska Salamander |
| Diamondback Terrapin | Longtail Salamander |
| Stripeneck Musk Turtle | Mole Salamander |
| Mimic Glass Lizard | Mudpuppy |
| Carolina Crawfish Frog | Neuse River Waterdog |
| Mountain Chorus Frog | Weller's Salamander |
| River Frog | Zigzag Salamander |

**REGULATIONS:**
    Persons wishing to collect reptiles and amphibians in this state should contact the Division of Wildlife Management for a scientific collection permit. As with most states, scientific collection permits will only be distributed to qualified individuals working on studies which will benefit the state's wildlife.
    Contact the Division for information about non-scientific collecting and possessing regulations.

# NORTH DAKOTA

Game and Fish Department
100 N. Bismark Expressway
Bismark, ND 58501-5095
Phone: (701) 221-6300

**ENDANGERED:**
North Dakota lists no Endangered or Threatened reptiles or amphibians.

**PERIPHERAL SPECIES:**
*(native species or subspecies with small or unknown populations whose breeding distribution or reproduction ability within the state of North Dakota is often severely limited by lack of suitable habitat or by climate)*
Smooth Softshell Turtle *(Trionyx muticus)*
False Map Turtle *(Graptemys pseudogeographica)*
Sagebrush Lizard *(Sceloporus graciosus)*
Prairie Skink *(Eumeces septentrionalis)*
Mudpuppy *(Necturus maculosus)*
Gray Treefrog *(Hyla versicolor)*

**REGULATIONS:**
Turtles are not to be taken without permit or contract from the Department. No person may engage in the commercial taking, trapping, or hooking of turtles without obtaining a permit from the Director, who may issue the permits at his discretion.
No person may engage in the taking of frogs for sale for human consumption or scientific purposes without obtaining a frog license from the Director. No person may buy, job, take on consignment, or ship frogs without obtaining the appropriate resident or non-resident commercial frog license. Except under certain circumstances, it is unlawful to take frogs on private land without written permission of the owner or operator of the land.

# OHIO

Department of Natural Resources
Division of Wildlife
1840 Belcher Dr.
Fountain Square, Bldg. G
Columbus, OH 43224-1329
Phone: (614) 265-6344
Fax: (614) 262-1143

**ENDANGERED:**
    Copperbelly Water Snake *(Nerodia erythrogaster neglecta)*
    Eastern Plains Garter Snake *(Thamnophis radix radix)*
    Timber Rattlesnake *(Crotalus horridus)*
    Hellbender *(Cryptobranchus alleganiensis)*
    Blue-spotted Salamander *(Ambystoma laterale)*
    Green Salamander *(Aneides aeneus)*
    Cave Salamander *(Eurycea lucifuga)*
    Eastern Spadefoot Toad *(Scaphiopus hoolbrooki)*

**THREATENED:**
    Lake Erie Water Snake *(Nerodia sipedon insularum)*
    Kirtland's Snake *(Clonophis kirtlandii)*

**SPECIES OF SPECIAL CONCERN:**

| | |
|---|---|
| Spotted Turtle | Rough Green Snake |
| Blanding's Turtle | Fox Snake |
| Coal Skink | Massasauga |
| Black Kingsnake | Four-toed Salamander |
| Common Garter Snake | Mud Salamander |

**REGULATIONS:**
    It is unlawful for any person to take frogs from May 1 to June 15 each year. It is unlawful for any person to take frogs by shooting, except they may be taken with a longbow and arrow. Daily limit is 10 frogs, and possession limit is 10 frogs, except those purchased out of state that are accompanied by a bill of lading. It is unlawful for any person to buy or sell frogs at any time. However, frogs that have been shipped from outside Ohio that are accompanied by a bill of lading may be bought and sold, and persons possessing a permit issued by the Department may sell frogs for propagation or stocking purposes when taking from ponds or lakes they own or lease. Persons selling frogs in this manner shall record each sale. Turtles may be bought and sold. It is unlawful for any person to take a turtle other than softshell, Snapping, and Midland Painted Turtles from property owned, controlled, or maintained by the Wildlife Division. It is unlawful for any person to take, possess, buy, sell, or transport lizards, snakes, and amphibians other than frogs, which have a daily limit and cannot be sold. Note that Pymatuning Lake has special requirements for taking frogs and turtles, and the open season and bag limits are different from the rest of the state (season is closed July 1 to October 31, possession limit is 15 frogs or 15 tadpoles, and 10 turtles (except Snappers—unlimited).

# OKLAHOMA

Department of Wildlife Conservation
P.O. Box 53465
Oklahoma City, OK 73152
Phone: (405) 521-3851

**ENDANGERED:**
    Oklahoma has no reptiles or amphibians listed as Endangered.

**THREATENED:**
    American Alligator *(Alligator mississippiensis)*

**SPECIES OF SPECIAL CONCERN:**
    Texas Horned Lizard *(Phrynosoma cornutum)*
    Desert Side-blotched Lizard *(Uta stansburiana)*
    Checkered Whiptail *(Cnemidophorus tesselatus)*
    Earless Lizard *(Holbrookia maculata)*
    Roundtail Horned Lizard *(Phrynosoma modestum)*
    Western Chicken Turtle *(Deirochelys reticularia)*
    Map Turtle *(Graptemys geographica)*
    Alligator Snapping Turtle *(Macroclemys temminckii)*
    Wandering Garter Snake *(Thamnophis elegans vagrans)*
    Gulf Crayfish Snake *(Regina rigida sinicola)*
    Sequoyah, Kiamichi, and Western Slimy Salamanders *(Plethodon* sp.)
    Rich Mountain Salamander *(Plethodon ouachitae)*
    Ozark Zigzag Salamander *(Plethodon dorsalis angusticlavius)*
    Four-toed Salamander *(Hemidactylium scutatum)*
    Grotto Salamander *(Typhlotriton spelaeus)*
    Oklahoma Salamander *(Eurycea tynerensis)*
    Cave Salamander *(Eurycea lucifuga)*
    Mole Salamander *(Ambystoma talpoideum)*

**REGULATIONS:**
    Species of special concern and the American Alligator have a statewide closed season. It is legal to take the Prairie Rattlesnake, Western Diamondback Rattlesnake, and Timber Rattlesnake from March 1 through June 30; there is no bag limit. All other reptiles have a year-round season; the bag limit is six per day or in possession. There is a statewide year-round open season on the following amphibians: *Rana* and *Pseudacris* (bag limit unlimited, except Bullfrogs have a bag limit of 15); *Necturus* and *Ambystoma* (bag limit unlimited) except *A. talpodium*, which is a species of special concern; all other amphibians have a bag limit of 4 per day or in possession of each species.
    Persons wishing to raise, breed, collect for hobby or commercial purposes, or otherwise possess, any lawfully obtained reptiles or amphibians must first obtain the appropriate: (a) Commercial Wildlife Breeders License; (b) Noncommercial Wildlife Breeders License; (c) Aquatic Culture License; (d) Commercial Fishing License. Anyone shipping or otherwise transporting wildlife into or out of the state must first apply for authorization on forms prescribed by the Department.

# *OREGON*

Department of Fish and Wildlife
2501 SW First Ave.
Portland, OR 97207
Phone: (503) 229-5400

**ENDANGERED:**
    Green Sea Turtle *(Chelonia mydas)*
    Leatherback Sea Turtle *(Dermochelys coriacea)*

**THREATENED:**
    Loggerhead Sea Turtle *(Caretta caretta)*
    Pacific Ridley Sea Turtle *(Lepidochelys olivacea)*

**PROTECTED WILDLIFE:**

| | |
|---|---|
| Cope's Giant Salamander | Northern Leopard Frog |
| Olympic Salamader | Spotted Frog |
| Clouded Salamander | Western Painted Turtle |
| Black Salamander | Western Pond Turtle |
| California Slender Salamander | Mojave Black-collared Lizard |
| Oregon Slender Salamander | Leopard Lizard |
| Del Norte Salamander | Short-horned Lizard |
| Larch Mountain Salamander | Desert Horned Lizard |
| Siskiyou Mountain Salamander | Sharptail Snake |
| Tailed Frog | Common Kingsnake |
| Red-legged Frog | California Mountain Kingsnake |
| Foothill Yellow-legged Frog | Western Ground Snake |
| Cascade Frog | |

**REGULATIONS:**

    The sale of wildlife is strictly regulated by Oregon law. ORS 498.002 states that "[E]xcept as the commission by rule may provide otherwise, no person shall purchase, sell, or exchange, or offer to purchase, sell or exchange any wildlife, or any part thereof."

    In addition to Oregon Department of Fish and Wildlife's statutes and rules, the Oregon Department of Agriculture has specific permit and holding requirements for the exotic animals listed under Oregon Revised Statutes 609.305 as well as importation requirements regarding disease control.

    Threatened or Endangered species may not be taken. Protected wildlife listed above are considered sensitive species and may not be taken except with a permit. Any other nongame wildlife species that is not Endangered, Threatened, or protected by statute or rule is nonprotected, and a permit is not required to hold that species. A Rehabilitation Holding Permit is required to rehabilitate amphibians or reptiles.

# PENNSYLVANIA

Fish & Boat Commission
Division of Fisheries Management
450 Robinson Lane
Bellefonte, PA 16823-9685
Phone: (814) 359-5110


**ENDANGERED:**
    Bog Turtle *(Clemmys muhlenbergi)*
    Massasauga Rattlesnake *(Sistrurus catenatus)*
    Kirtland's Snake *(Clonophis kirtlandii)*
    New Jersey Chorus Frog *(Pseudacris feriarum kalmi)*
    Coastal Plain Leopard Frog *(Rana utricularia)*
    Eastern Mud Salamander *(Pseudotriton montanus montanus)*

**THREATENED:**
    Red-bellied Turtle *(Pseudemys rubriventris)*
    Rough Green Snake *(Opheodrys aesticus)*
    Green Salamander *(Aneides aeneus)*

**SPECIES OF SPECIAL CONCERN:**
    Blanding's Turtle *(Emydoidea blandingii)*
    Timber Rattlesnake *(Crotalus horridus)*
    Broadhead Skink *(Eumeces laticeps)*

**REGULATIONS:**
    Endangered and Threatened species may not be taken, killed, possessed, imported to or exported from Pennsylvania without a special permit. It is unlawful to take, catch, kill, or possess a Timber Rattlesnake in a hunt for such creature without first procuring the required permit. The sponsors of organized reptile and amphibian hunts must also apply for a permit. It is unlawful to take, catch, or kill any amphibian or reptile (except the Snapping Turtle) for the purpose of selling. Contact the Herpetology and Endangered Species Coordinator at the above address for permit information.

    A fishing license is required by persons age 16 and over to take all reptiles and amphibians. The season for Bullfrogs and Green Frogs is July 1 to October 31; daily limit is 15 of each species, and possession limit is unlimited. It is illegal to use artificial light to take frogs at night. With a permit, Timber Rattlesnake season is the second Saturday in June to July 31; daily limit is one and possession limit is one. The Eastern Mud Salamander, Blanding's Turtle, and Broadhead Skink have no open season. Tadpoles have no closed season; daily limit and possession limit are 15. The Snapping Turtle has no closed season; daily limit and possession limit are unlimited. Finally, each species of Pennsylvania amphibians and reptiles not listed in the paragraph (excluding Endangered and Threatened species) has no closed season, with a daily limit and possession limit of two.

    It is unlawful to introduce any species of reptile or amphibian into the natural environment of Pennsylvania if that species is not native to or does not occur within this state.

# PUERTO RICO

Department of Natural Resources
P.O. Box 5887
Puerta de Tierra
San Juan, PR 00906
Phone: (809) 724-8774

**ENDANGERED:**
      Roosevelt Anole *(Anolis roosevelti)*
      Monito Reef Gecko *(Sphaerodactylus micropithecus)*
      Green Sea Turtle *(Chelonia mydas)*
      Leatherback Sea Turtle *(Dermochelys coriacea)*
      Hawksbill Sea Turtle *(Eretmochelys imbricata)*
      Kemp's Ridley Sea Turtle *(Lepidochelys kempii)*
      Puerto Rican Boa *(Epicrates inornatus)*
      Mona Island Boa *(Epicrates monensis grantii)*

**THREATENED:**
      *Bufo lemur*
      *Eleutherodactylus eneidae*
      *Eleutherodactylus jasperi*
      *Eleutherodactylus karlschmidti*
      *Anolis cooki*
      *Cyclura stejnegeri*
      *Mabuya mabouya sloanii*
      *Caretta caretta*
      *Epicrates monensis monensis*

**REGULATIONS:**
      Endangered or Threatened species may not be taken. A scientific collection permit is available for recognized scientists or educators, through the Department of Natural Resources. Permitees must provide detailed information on their background, the study, a description of the species (ages, sexes, etc.) which will be collected, how many individuals were taken, etc.

# RHODE ISLAND

Department of Environmental Management
Division of Fish and Wildlife
Field Headquarters, Box 218
West Kingston, RI 02892
Phone: (401) 789-0281

**ENDANGERED:**
Except for the federally listed sea turtles, Rhode Island has no amphibian or reptile species listed as Endangered.

**THREATENED:**
Eastern Spadefoot Toad *(Scaphiopus holbrooki)*
Northern Diamondback Terrapin *(Malaclemys terrapin)*

**SPECIES OF SPECIAL CONCERN & STATE INTEREST:**

| | |
|---|---|
| Northern Spring Salamander | Eastern Hognose Snake |
| Marbled Salamander | Eastern Worm Snake |
| Northern Leopard Frog | Eastern Ribbon Snake |
| Wood Turtle | Black Rat Snake |

**REGULATIONS:**
Common Snapping Turtles may be harvested for sale according to the regulations of the Rhode Island Division of Fish and Wildlife; a permit is required. Individuals of any native reptile or amphibian species may not be collected for sale without special permit through the Division. Scientific collection is allowed through permit from the Division. The following species may not be possessed without a permit: Marbled Salamander, Timber Rattlesnake, Diamondback Terrapin, Wood Turtle, and Eastern Box Turtle. Except for previously listed species, individuals of native species may be possessed without permit. There are currently no numerical or temporal limits to possession.

No individuals of non-native species listed as federally Endangered or Threatened species according to the Endangered Species Act, and no individuals regulated by other national and international treaties (C.I.T.E.S.) may be imported without permit from the U.S. Fish and Wildlife Service.

Rhode Island has rules and regulations pertaining to the importation of captive wild and exotic animals. These cover animal importation as it relates to public health issues and establish humane husbandry criteria for captive animals. Importation of amphibians and reptiles thought to represent a threat to humans, native wildlife, or domestic animals is not allowed without permit from the state veterinarian, Division of Agriculture. These regulations are currently being reorganized. Species prohibited from importation include (but are not limited to): Order Crocodilia, Family Helodermatidae, Family Varanidae (in part, large monitor lizards), Family Boidae (in part, large constricting snakes), Family Elapidae, Family Hydrophiidae, Family Viperidae, and Family Crotalidae.

# SOUTH CAROLINA

Wildlife & Marine Resources Department
P.O. Box 167
Columbia, SC 29202
Phone: (803) 734-3886

**ENDANGERED:**
    Flatwoods Salamander *(Ambystoma cingulatum)*
    Webster's Salamander *(Plethodon websteri)*
    Gopher Tortoise *(Gopherus polyphemus)*
    Green Sea Turtle *(Chelonia mydas)*
    Eastern Indigo Snake *(Drymarchon corais couperi)*

**THREATENED:**
    Bog Turtle *(Clemmys muhlenbergi)*
    Southern Coal Skink *(Eumeces anthracinus pluvialis)*
    Pine Barrens Treefrog *(Hyla andersonii)*
    Dwarf Siren *(Pseudobranchus striatus)*

**SPECIES OF SPECIAL CONCERN:**
    Eastern Tiger Salamander *(Ambystoma tigrinum tigrinum)*
    Gopher Frog *(Rana capito)*
    Florida Softshell Turtle *(Trionyx ferox)*
    Striped Mud Turtle *(Kinosternon baurii)*

**REGULATIONS:**
    There are no laws prohibiting the possession of any but state or federally listed herps. A collecting permit is required only if one is collecting for purposes of science or for propagation. According to Stephen H. Bennett, Inventory Coordinator of the Heritage Trust Program, this is essentially backwards of what the regulation should be, but sometimes that happens in bureaucracies. The Heritage Trust Program is discussing ways in which these regulations can be strengthened, as "It would be good to protect species sensitive to overcollection due to commercial demand without stopping folks from enjoying some hobby snake collecting."

**NOTES:**
    The Bog Turtle and the Striped Mud Turtle have only recently been reported from South Carolina. The Striped Mud Turtle may be fairly common, yet commonly misidentified as the Eastern Mud Turtle, as individuals in South Carolina don't always have good carapacial or head stripes. The Bog Turtle is truly rare in the state, as only five specimens have been reported despite extensive survey efforts.

# SOUTH DAKOTA

Department of Game, Fish and Parks
Division of Wildlife
523 E. Capitol
Pierre, SD 57501
Phone: (605) 773-3485

**ENDANGERED:**
   No reptile or amphibian species in South Dakota are considered Endangered.

**THREATENED:**
   Spiny Softshell Turtle *(Apalone spinifera)*
   Blanding's Turtle *(Emydoidea blandingii)*
   False Map Turtle *(Graptemys pseudogeographica)*
   Eastern Hognose Snake *(Heterodon platirhinos)*
   Northern Redbelly Snake *(Storeria o. occipitomaculata)*
   Lined Snake *(Tropidoclonion lineatum)*

**REGULATIONS:**
   Most of the herp species (including snakes, lizards, frogs, and salamanders are considered to be bait or biological specimens by South Dakota law. Licensed anglers may take bait for non-commercial purposes with a limit of 12 dozen of any combination of species. Non-residents may not take or collect any of these bait species for commercial purposes, however they could buy bait from a resident and take them out of state for commercial purposes if they first obtain an Export Permit from the Game, Fish, and Parks at a cost of $400 annually. South Dakota licenses are not transferable. The license year is from January 1 through December 31, with annual, five-day, or 24-hour licenses and/or stamps available.
   Licensed anglers may take Bullfrogs from May 1 through October 15. The limit is 15 daily, 30 in possession. Licensed anglers may take all other frogs for bait all year. Persons wanting to buy, sell, or transport frogs for commercial purposes, contact the Licensing Office.
   Licensed anglers may take turtles from January 1 to December 31 by hook and line, legal minnow seines, gaff hooks, spears, or legal turtle trap. The limit for Snapping Turtles is two daily, four in possession. There is a 12 dozen limit on the taking of other common turtles. Turtle traps with mesh less than four-inches-square must have an opening of at least six inches in diameter leading from it or an entrance opening suspended at or above the water level. Traps must be clearly marked with the owner's name and address. It is illegal to buy, sell, barter, or trade Snapping Turtles taken in South Dakota or to export Snapping Turtles for any purpose other than personal consumption.
   Endangered/Threatened species must be released when taken. Some municipalities have their own ordinances concerning the possession of various amphibian or reptile species, which they may consider as harmful or injurious. At the present time the Department is involved in a planning process, looking at the various laws and regulations they enforce. Contact the Department for recent developments.

# TENNESSEE

Wildlife Resources Agency
Ellington Agricultural Center
P.O. Box 40747
Nashville, TN 37204-0747
Phone: (615) 781-6500

**ENDANGERED:**
There are no Endangered reptiles or amphibians listed in Tennessee.

**THREATENED:**
Tennessee Cave Salamander *(Gyrinophilus palleucus)*
Bog Turtle *(Clemmys muhlenbergi)*
Northern Pine Snake *(Pituophis m. melanoleucus)*
Western Pigmy Rattlesnake *(Sistrurus miliarius streckeri)*

**WILDLIFE IN NEED OF MANAGEMENT:**

| | |
|---|---|
| Hellbender | Alligator Snapping Turtle |
| Green Salamander | Cumberland Slider |
| Black Mountain Dusky Salamander | Green Anole |
| Four-toed Salamander | Six-lined Racerunner |
| Barking Treefrog | Eastern Slender Glass Lizard |
| Green Water Snake | |

**REGULATIONS:**
It is unlawful for any person to take, harass, or destroy wildlife listed as Threatened or Endangered. It is unlawful for any person to knowingly destroy the habitat of wildlife in need of management. The Executive Director of the Tennessee Wildlife Resources Agency may permit the taking, possession, transportation, exportation or shipment of species or subspecies on the list of wildlife in need of management for scientific or educational purposes, for propagation in captivity or for other purposes to benefit the species.

# TEXAS

Parks and Wildlife Department
4200 Smith School Rd.
Austin, TX 78744
Phone: (512) 389-4800

**ENDANGERED**:
- Chihuachuan Mud Turtle *(Kinosternon hirtipes murrayi)*
- Loggerhead Sea Turtle *(Caretta caretta)*
- Hawksbill Sea Turtle *(Eretmochelys imbricata)*
- Kemp's Ridley Sea Turtle *(Lepidochelys kempii)*
- Leatherback Sea Turtle *(Dermochelys coriacea)*
- Louisiana Pine Snake *(Pituophis melanoleucus ruthveni)*
- Northern Cat-eyed Snake *(Leptodeira septentrionalis)*
- Smooth Green Snake *(Opheodrys vernalis)*
- Concho Water Snake *(Nerodia harteri paucimaculata)*
- Speckled Racer *(Drymobius margaritiferus)*
- Houston Toad *(Bufo houstonensis)*
- White-lipped Frog *(Leptodactylus labialis)*
- Black-spotted Newt *(Notophthalmus meridionalis)*
- Rio Grande Lesser Siren *(Siren intermedia texana)*
- Texas Blind Salamander *(Typhlomolge rathbuni)*
- Blanco Blind Salamander *(Typhlomolge robusta)*

**THREATENED**:

| | |
|---|---|
| Alligator Snapping Turtle | Texas Lyre Snake |
| Texas Tortoise | Texas Horned Lizard |
| Green Sea Turtle | Mountain Short-horned Lizard |
| Northern Scarlet Snake | Reticulate Collared Lizard |
| Texas Scarlet Snake | Mexican Treefrog |
| Black-striped Snake | Sheep Frog |
| Indigo Snake | Mexican Burrowing Toad |
| Timber Rattlesnake | San Marcos Salamander |
| Brazos Water Snake | Comal Blind Salamander |
| Big Bend Blackhead Snake | Cascade Caverns Salamander |

**REGULATIONS**:

All wildlife classified as Endangered are illegal to possess, take, or transport these species for zoological gardens or scientific purposes or to take or transport Endangered species from their natural habitat for propagation for commercial purposes without a special permit. No person may possess, sell, distribute, offer or advertise for sale Endangered species unless the wildlife have been lawfully born and raised in captivity for commercial purposes under the provisions of this chapter.

The Department may issue permits for the taking, possession, transportation, sale, or exportation of a nongame species of wildlife if necessary to properly manage that species.

# UTAH

Wildlife Resources
Nongame Wildlife
1596 West North Temple
Salt Lake City, UT 84116-3195
Phone: (801) 538-4700

**ENDANGERED:**
   Lowland Leopard Frog *(Rana yavapaiensis)*

**THREATENED:**
   Southwestern Toad *(Bufo microscaphus)*
   Spotted Frog *(Rana pretiosa)*
   Gila Monster *(Heloderma suspectum)*
   Desert Tortoise *(Gopherus agassizii)*

**SPECIES OF SPECIAL CONCERN:**

| | |
|---|---|
| Western Toad | Glossy Snake |
| Northern Leopard Frog | Utah Mountain Kingsnake |
| Pacific Chorus Frog | Utah Milk Snake |
| Chuckwalla | Lyre Snake |
| Desert Night Lizard | Western Blind Snake |
| Desert Iguana | Western Patchnose Snake |
| Zebra-tailed Lizard | Speckled Rattlesnake |
| Western Banded Gecko | Mojave Rattlesnake |
| Great Plains Rat Snake | Sidewinder |

**REGULATIONS:**
   Utah has detailed rules on the "Collection, Importation, Transportation and Subsequent Possession of Zoological Animals" which covers amphibians and reptiles. Indiscriminate killing of ALL amphibians and reptiles is prohibited. Except as otherwise provided, a certificate of registration will not be issued for collection or killing of prohibited amphibians. Prohibited amphibian—Spotted Frog. A certificate of registration is required for collection or killing of controlled amphibians—Arizona Toad and Pacific Treefrog. The Bullfrog, Green Frog, and Arizona Tiger Salamander are noncontrolled.
   It is unlawful to disturb the den of any reptile or to kill, capture, or harass any reptile within 100 yards of a reptile den without first obtaining a certificate of registration from Wildlife Resources. Prohibited reptiles—Chuckwalla, Desert Iguana, Gila Monster, Desert Glossy Snake, Sonoran Lyre Snake, Utah Milk Snake, Utah Mountain Kingsnake, Mojave Rattlesnake, Sidewinder, Speckled Rattlesnake, Western Blind Snake, Desert Tortoise. Controlled reptiles—Common Zebra-tailed Lizard, Desert Night Lizard, Common Night Lizard, Utah Night Lizard, Utah Banded Gecko, California Kingsnake, Great Plains Rat Snake, Mojave Patchnose Snake, Western Rattlesnake.

# VERMONT

Fish & Wildlife Department
103 S. Main St.
Waterbury, VT 05676
Phone: (802) 241-3700

**ENDANGERED:**
   Striped Chorus Frog *(Pseudacris triseriata)*
   Timber Rattlesnake *(Crotalus horridus)*
   Five-lined Skink *(Eumeces fasciatus)*

**THREATENED:**
   Spiny Softshell Turtle *(Apalone spinifera)*
   Spotted Turtle *(Clemmys guttata)*

**REGULATIONS:**
   The commissioner may issue permits to a properly accredited person or educational institution permitting the holder thereof to collect birds, their nests and eggs, and fish and wild animals or parts thereof, for public scientific research or educational purposes of the institution. An applicant for a permit shall present to the commissioner written testimonials from two fisheries and wildlife biologists, certifying to the fitness of the applicant. The permit shall expire on December 31 in the year in which it is issued and shall be revoked by the commissioner upon proof that the holder of the permit has violated its provisions. In addition, the commissioner may issue a permit to an individual which allows the holder to collect fish and wild animals for the purposes of using them as subjects of art or photography.
   A person shall not throw or cast the rays of a spotlight, jack, or other artificial light on any highway, or any field, woodland, or forest, for the purpose of spotting, locating or taking any wild animal, except that a light may be used to take skunks and raccoons in accordance with rules of the board.
   Dealers who wish to import herps into Vermont must fill out an application for an Importation Permit.

# VIRGIN ISLANDS

Department of Planning and Natural Resources
Division of Fish and Wildlife
6291 Estate Nazareth, 101
St. Thomas, VI 00802-1104
Phone: (809) 775-6762

**ENDANGERED:**
    Hawksbill Sea Turtle *(Eretmochelys imbricata)*
    Leatherback Sea Turtle *(Dermochelys coriacea)*
    Virgin Islands Tree Boa *(Epicrates monensis granti)*
    St. X Ground Lizard *(Ameiva polops)*
    Slipperyback Skink *(Mabuya mabouia)*

**THREATENED:**
    Green Turtle *(Chelonia mydas)*

**REGULATIONS:**
    No person may take, catch, possess, injure, harass, kill, or attempt to take, catch, possess, injure, harass or kill, or sell or offer for sale, or transport to export, whether or not for sale, any indigenous species—except persons holding valid fishing or hunting licenses, scientific or aquarium collecting permits, or indigenous species retention permits.

    No person may take, catch, or possess, or attempt to take, catch or possess, any specimen of an Endangered or Threatened species unless such person holds a valid collecting permit from the Federal Government in the case of a Federally listed species, or a Territorial permit in the case of an exclusively territorially listed species. No person may ship, transport, or export any specimen of an Endangered or Threatened species, or parts or produce thereof, whether for sale or not, unless such person holds a valid Federal or Territorial permit.

    It shall be unlawful for any person to import or introduce, or cause the importation or introduction to the Virgin Islands of any species of plant or animal which does not naturally occur in the Territory without the express written permission of the Commissioner. No person may harass, injure or kill, or attempt to do the same, or sell or offer for sale any specimen, or parts or produce of such a specimen, of an Endangered or Threatened species. No person may disturb, damage or remove the nest, or contents thereof, of any indigenous, Endangered species.

# VIRGINIA

Department of Game and Inland Fisheries
4010 W. Broad St.
Richmond, VA 23230-1104
Phone: (804) 367-1000
Fax: (804) 367-9147

**ENDANGERED:**
- Eastern Tiger Salamander *(Ambystoma tigrinum)*
- Canebrake Rattlesnake *(Crotalus horridus atriccaudatus)*
- Bog Turtle *(Clemmys muhlenbergi)*
- Chicken Turtle *(Deirochelys reticularia)*

**THREATENED:**
- Mabee's Salamander *(Ambystoma mabeei)*
- Barking Treefrog *(Hyla gratiosa)*
- Eastern Glass Lizard *(Ophisaurus ventralis)*
- Wood Turtle *(Clemmys insculpta)*

**SPECIES OF SPECIAL CONCERN:**

| | |
|---|---|
| Carpenter Frog | Peaks of Otter Salamander |
| Oak Toad | Pigmy Salamander |
| Cow Knob Salamander | Shovelnose Salamander |
| Eastern Hellbender | Spotbelly Salamander |
| Mole Salamander | Mountain Earth Snake |

**REGULATIONS:**

It is unlawful to capture and possess live for private use and not for sale no more than five individuals of any single native or naturalized species of amphibian and reptile. Fish bait (includes salamanders) possession limit is 50 individuals in aggregate. The daily limit for Bullfrogs and Snapping Turtles shall be 15 and Bullfrogs and Snapping Turtles may not be taken from the banks or waters of designated stocked trout waters. It is unlawful to take Snapping Turtles for sale.

A special permit is required and may be issued by the Department, if consistent with the fish and wildlife management program, to import, possess, or sell those non-native (exotic) animals listed below: Giant Marine Toad, African Clawed Frog, Barred Tiger Salamander, Gray Tiger Salamander, Blotched Tiger Salamander, Brown Tree Snake, and all species of alligators, caimans, crocodiles, and gavials. All other non-native animals may be possessed and sold; provided, that such animals shall be subject to all applicable local, state, and federal laws and regulations, including those that apply to Threatened/Endangered species, and further provided, that such animals shall not be liberated within the Commonwealth.

# WASHINGTON

Department of Wildlife
600 Capitol Way North
Olympia, WA 98501-1091
Phone: (206) 753-5700

**ENDANGERED:**
    Leatherback Sea Turtle *(Dermochelys coriacea)*
    Western Pond Turtle *(Clemmys marmorata)*

**PROTECTED:**
    Painted Turtle *(Chrysemys picta)*
    Pond Slider *(Pseudemys scripta)*
    Olive Ridley Sea Turtle *(Lepidochelys olivacea)*
    Green Sea Turtle *(Chelonia mydas)*
    Loggerhead Sea Turtle *(Caretta caretta)*
    Larch Mountain Salamander *(Plethodon larselli)*

**REGULATIONS:**
    The regulations in the Wildlife Code of the State of Washington are based on species' classifications that are legally defined. Reptiles and amphibians fall into a number of classifications: Endangered, Protected (which includes the subcategories Threatened, Sensitive, and Other), Deleterious Exotic, Unclassified Wildlife, and Non-wildlife (species whose members do not exist in a wild state in Washington). The regulations which are pertinent depend upon the classification of the species of interest.
    The Bullfrog is considered a game animal. The African Clawed Frog is classified as Deleterious Exotic Wildlife. Non-wildlife species do not receive regulatory protection of any kind.

# WEST VIRGINIA

Division of Natural Resources
Operations Center
P.O. Box 67
Elkins, WV 26241-0067
Phone: (304) 637-0245
Fax: (304) 637-0250

**THREATENED:**
    Cheat Mountain Salamander *(Plethodon nettingi)*

**SPECIES OF SPECIAL CONCERN:**

| | |
|---|---|
| Eastern Hellbender | Ouachita Map Turtle |
| Streamside Salamander | Red-eared Slider |
| Jefferson Salamander | Eastern River Cooter |
| Smallmouth Salamander | Hieroglyphic Turtle |
| Blackbelly Salamander | Redbelly Turtle |
| Cave Salamander | Midland Smooth Softshell Turtle |
| White-spotted Salamander | Ground Skink |
| West Virginia Spring Salamander | Northern Coal Skink |
| Green Salamander | Broadhead Skink |
| Eastern Spadefoot Toad | Eastern Ribbon Snake |
| Cricket Frog | Mountain Earth Snake |
| Upland Chorus Frog | Corn Snake |
| Northern Leopard Frog | Northern Pine Snake |
| Map Turtle | Eastern Kingsnake |
| Spotted Turtle | Timber Rattlesnake |
| Wood Turtle | Eastern Hognose Snake |

**REGULATIONS:**
    West Virginia does not have a state law regarding Endangered and Threatened species. The Natural Heritage Program does monitor rare herptiles in the state.
    Possession of reptiles and amphibians requires a state fishing license. There is no commercial collecting permit in West Virginia. Collections can be made by acquiring a Scientific Collecting Permit. All permits will be reviewed by state wildlife biologists and a determination made if the permit will be issued. Permit requests should be made to Barbara Sargent, Scientific Collecting Permits, P.O. Box 67, Elkins, WV 26241.

# WISCONSIN

Department of Natural Resources
Box 7921
Madison, WI 53707
Phone: (608) 266-2621

**ENDANGERED:**
    Blanchard's Cricket Frog *(Acris crepitans)*
    Slender Glass Lizard *(Ophisaurus attenuatus)*
    Queen Snake *(Regina septemvittata)*
    Massasauga *(Sistrurus catenatus)*
    Western Ribbon Snake *(Thamnophis proximus)*
    Northern Ribbon Snake *(Thamnophis sauritus)*
    Ornate Box Turtle *(Terrapene ornata)*

**THREATENED:**
    Wood Turtle *(Clemmys insculpta)*
    Blanding's Turtle *(Emydoidea blandingi)*

**SPECIES OF SPECIAL CONCERN:**
    Timber Rattlesnake *(Crotalus horridus)*
    Butler's Garter Snake *(Thamnophis butleri)*

**REGULATIONS:**
    Any person may sell, buy, barter or trade; or, offer to sell, buy, barter or trade; or, have in possession or under control for the purpose of trade any amphibian or reptile, with the following exceptions:

    1. No person may take, transport, possess, process or sell within Wisconsin any wild animal specified by the DNR as an Endangered or Threatened species.

    2. Frogs may be taken during open season from the Saturday nearest May 1 to December 31. Frogs, or any part of, may not be used as bait in the Mississippi River. There is no open season for Bullfrogs in Jefferson and Walworth counties.

    3. Turtles may be taken during open season from areas except the Wisconsin-Minnesota and Wisconsin-Iowa (Mississippi River) boundary waters from June 16-April 30 by trapping or hooking. Turtles may be taken when an individual is in compliance with: licensing; regulation of trap construction; transportation and with the regulations of neighboring states. On the Mississippi River, turtles may be taken year-round. Snapping Turtles must have a minimum carapace length of 10 inches.

    In total not more than five hoop net turtle traps may be used on Wisconsin Inland and Wisconsin-Minnesota-Iowa boundary waters. The traps must have a mesh of not less than an eight-inch stretch measure (six-inch for Wisconsin-Minnesota-Iowa boundary waters). No person may operate more than 40 traps on any boundary waters. Except for hoop nets and boundary waters listed above, there is no limit for other types of turtle traps on inland waters. Turtle traps must be raised and the contents removed at least once each day following the day set in inland waters.

# WYOMING

Game and Fish Department
5400 Bishop Blvd.
Cheyenne, WY 82006-0001
Fax: (307) 777-4610




**ENDANGERED:**
Wyoming has no official state herptile Endangered/Threatened list.

**S1—EXTREMELY RARE, CRITICALLY IMPERILED:**
Wyoming Toad *(Bufo hemiophrys baxteri)*

**S2—RARE, IMPERILED:**
Boreal Toad *(Bufo boreas)*
Boreal Western Toad *(Bufo boreas boreas)*
Wood Frog *(Rana sylvatica)*
Northern Earless Lizard *(Holbrookia maculata)*
Northern Tree Lizard *(Urosaurus ornatus)*
Rubber Boa *(Charina bottae)*
Black Hill Redbelly Snake *(Storeria occipitomaculata pahasapae)*

**S3—RARE, UNCOMMON**
Northern Leopard Frog *(Rana pipiens)*
Spotted Frog *(Rana pretiosa)*
Ornate Box Turtle *(Terrapene ornata)*
Northern Plateau Lizard *(Sceloporus undulatus elongatus)*
Red-lipped Prairie Lizard *(Sceloporus undulatus erythrocheilus)*
Prairie Lined Racerunner *(Cnemidophorus sexlineatus)*
Pale Milke Snake *(Lampropeltis triangulum)*
Smooth Green Snake *(Opheodrys vernalis)*
Western Smooth Green Snake *(Opheodrys vernalis blanchardi)*
Eastern Smooth Green Snake *(Opheodrys vernalis vernalis)*
Midget Faded Rattlesnake *(Crotalus viridis concolor)*

**REGULATIONS:**
The state of Wyoming has no official herptile Endangered/Threatened list. The above list was prepared by the Wyoming branch of the Nature Conservancy.

Currently, Wyoming has no regulations on the taking of reptile and amphibian species, although they do follow the Federal list mandated under the Endangered Species Act. In recent years, this has been recognized as a situation that needs changing. This has in part been prompted by a couple of instances where large numbers of certain species were collected by individuals and shipped out of state for commercial purposes. Department officials expect to see some form of regulations on reptiles and amphibians in the near future.

# C.I.T.E.S. & FEDERAL LISTS

## CLASS AMPHIBIA

| | | CITES | E/T |
|---|---|---|---|
| **Anura—Atelopidae** | | | |
| *Atelopus varius zeteki* | Panamanian Golden Frog | I | E |
| **Bufonidae** | | | |
| *Bufo hemiophrys baxteri* | Wyoming Toad | | E |
| *Bufo houstonensis* | Houston Toad | | E |
| *Bufo periglenes* | Monte Verde Toad (Costa Rica) | III | E |
| *Bufo retiformis* | Sonoran Green Toad | II | |
| *Bufo superciliaris* | Cameroon Toad | I | E |
| *Nectophrynoides occidentalis* | African Viviparous Toad | I | E |
| *Nectophrynoides tornieri* | Tanzania Viviparous Toad | I | E |
| *Nectophrynoides vivipara* | Viviparous Toad | I | E |
| *Peltophryne lemur* | Puerto Rican Crested Toad | | T |
| **Dendrobatidae** | | | |
| *Dendrobates* (all species) | poison arrow frogs | II | |
| *Phyllobates* (all species) | poison arrow frogs | II | |
| **Discoglossidae** | | | |
| *Discoglossus nigriventer* | Israel Painted Frog | | T |
| **Leiopelmatidae** | | | |
| *Leiopelma hamiltoni* | Stephen Island Frog | | E |
| **Leptodactylidae** | | | |
| *Eleutherodactylus jasperi* | Golden Coqui | | T |
| *Rheobatrachus silus* | Platypus Frog | II | |
| **Microhylidae** | | | |
| *Dyscophus antongilii* | Tomato Frog | I | |

55

|  |  | CITES | E/T |
|---|---|---|---|
| **Ranidae** | | | |
| *Rana hexadactyla* | Asian Bullfrog | II | |
| *Rana tigerina* | Indian Bullfrog | II | |
| **Caudata—Ambystomatidae** | | | |
| *Ambystoma dumerilii* | Lake Patzcuaro Salamander | II | |
| *Ambystoma macrodactylum croceum* | Santa Cruz Long-toed Salamander | | E |
| *Ambystoma mexicanum* | Axolotl | II | |
| **Cryptobranchidae** | | | |
| *Andrias davidianus japonicus* | Japanese Giant Salamander | I | E |
| *Andrias davidianus davidianus* | Chinese Giant Salamander | I | E |
| **Plethodontidae** | | | |
| *Batrachoseps aridus* | Desert Slender Salamander | | E |
| *Eurycea nana* | San Marcos Salamander | | T |
| *Phaeognathus hubrichti* | Red Hills Salamander | | T |
| *Plethodon nettingi* | Cheat Mountain Salamander | | T |
| *Plethodon shennadoah* | Shenandoah Salamander | | E |
| *Typhlomolge rathbuni* | Texas Blind Salamander | | E |

## CLASS REPTILIA

**Crocodylia—Alligatoridae**

|  |  | CITES | E/T |
|---|---|---|---|
| *Alligator mississippiensis* | American Alligator | II | T |
| *Alligator sinensis* | Chinese Alligator | I | E |
| *Caiman crocodilus crocodilus* | Common Caiman | II | |
| *Caiman crocodilus apaporiensis* | Apaporis River Caiman | I | E |
| *Caiman crocodilus fuscus* | Brown Caiman | II | |
| *Caiman crocodilus yacare* | Yacare Caiman | II | E |
| *Caiman latirostris* | Broad-snouted Spectacled Caiman | I | E |
| *Melanosuchus niger* | Black Caiman | I | E |
| *Paleosuchus palpebrosus* | Dwarf Caiman | II | |
| *Paleosuchus trigonotus* | Smooth-fronted Caiman | II | |

**Crocodylidae**

|  |  | CITES | E/T |
|---|---|---|---|
| *Crocodylus acutus* | American Crocodile | I | E |
| *Crocodylus cataphractus* | African Slender-snouted Crocodile | I | E |
| *Crocodylus intermedius* | Orinoco Crocodile | I | E |
| *Crocodylus johnsoni* | Johnson's Crocodile | II | |
| *Crocodylus moreletii* | Morelet's Crocodile | I | E |
| *Crocodylus niloticus* | Nile Crocodile—except below | I | E |
|  | (Madagascar, Somalia, South Africa, Uganda) | II | E |
|  | (Zimbabwe) | II | T |
|  | (Botswana, Ethiopia, Mozambique, Tanzania, Zambia) | II | E |
| *Crocodylus novaeguineae* | Freshwater Crocodile | II | |
| *Crocodylus novaeguineae mindorensis* | Philippine Crocodile | I | E |
| *Crocodylus palustris kimbula* | Ceylon Mugger Crocodile | I | E |

|  |  | CITES | E/T |
|---|---|---|---|
| *Crocodylus palustris palustris* | Mugger Crocodile | I | E |
| *Crocodylus porosus* | Saltwater Crocodile—except below | I | E |
|  | (Australia, Indonesia) | II |  |
|  | (Papua New Guinea) | II |  |
| *Crocodylus rhombifer* | Cuban Crocodile | I | E |
| *Crocodylus siamensis* | Siamese Crocodile | I | E |
| *Osteolaemus tetraspis* | Dwarf Crocodile | I |  |
| *Osteolaemus tetraspis osborni* | Congo Dwarf Crocodile | I | E |
| *Osteolaemus tetraspis tetraspis* | African Dwarf Crocodile | I | E |
| *Tomistoma schlegelii* | False Gavial | I | E |

## Gavialidae

| *Gavialis gangeticus* | Gavial | I | E |
|---|---|---|---|

## Rhynchocephalia—Sphenodontidae

| *Sphenodon punctatus* | Tuatara | I | E |
|---|---|---|---|

## Squamata—Agamidae

| *Uromastyx* (all species) | spiny-tailed lizards | II |  |
|---|---|---|---|

### Agavaceae

| *Dracaena guianensis* | Caiman Lizard | II |  |
|---|---|---|---|
| *Dracaena paraguayensis* | Paraguay Caiman Lizard | II |  |

### Boidae

| *Acrantophis dumerili* | Dumeril's Madagascan Boa | I |  |
|---|---|---|---|
| *Acrantophis madagascariensis* | Madagascan Boa | I |  |
| *Aspidites melanocephalus* | Blackheaded Python | II |  |
| *Aspidites ramsayi* | Woma | II |  |
| *Boa constrictor* | Boa Constrictor | II |  |
| *Boa constrictor occidentalis* | Argentine Boa Constrictor | I |  |
| *Bolyeria multocarinata* | Round Island Burrowing Boa | I | E |
| *Calabaria reinhardtii* | African Burrowing Python | II |  |
| *Candoia aspera* | New Guinea Viper Boa | II |  |
| *Candoia bibroni* | Pacific Island Boa | II |  |
| *Candoia carinata* | Solomon Islands Ground Boa | II |  |
| *Casarea dussumieri* | Round Island Boa | I | E |
| *Charina bottae* | Rubber Boa | II |  |
| *Chondropython viridis* | Green Tree Python | II |  |
| *Corallus annulatus* | Annulated Boa | II |  |
| *Corallus caninus* | Emerald Tree Boa | II |  |
| *Corallus eydris* | Amazon Tree Boa | II |  |
| *Epicrates angulifer* | Cuban Boa | II |  |
| *Epicrates cenchria* | Rainbow Boa | II |  |
| *Epicrates chrysogaster* | Turk's Island Boa | II |  |
| *Epicrates exsul* | Abaco Island Boa | II |  |
| *Epicrates fordii* | Ford's Boa | II |  |
| *Epicrates gracilis* | Vine Boa | II |  |
| *Epicrates inornatus* | Puerto Rican Boa | I | E |
| *Epicrates monensis monensis* | Mona Island Boa | I | T |
| *Epicrates monensis granti* | Virgin Islands Tree Boa | I | E |
| *Epicrates striatus* | Haitian Boa | II |  |

|  |  | CITES | E/T |
|---|---|---|---|
| *Epicrates subflavus* | Jamaican Boa | I | E |
| *Eryx colubrinus* | Egyptian Sand Boa | II | |
| *Eryx conicus* | Rough-scaled Sand Boa | II | |
| *Eryx elegans* | Central Asia Sand Boa | II | |
| *Eryx jaculus* | Caucasian Sand Boa | II | |
| *Eryx jayakari* | Arabian Sand Boa | II | |
| *Eryx johnii* | Smooth Scaled Sand Boa | II | |
| *Eryx miliaris* | Dwarf Sand Boa | II | |
| *Eryx muelleri* | Mueller's Sand Boa | II | |
| *Eryx nogaiorum* | Black Sand Boa | II | |
| *Eryx somalicus* | Somalian Sand Boa | II | |
| *Eryx tataricus* | Tartar Sand Boa | II | |
| *Eunectes barbouri* | Barbour's Anaconda | II | |
| *Eunectes deschauenseei* | Dark-spotted Anaconda | II | |
| *Eunectes murinus* | Green Anaconda | II | |
| *Eunectes notaeus* | Yellow Anaconda | II | |
| *Exiliboa placata* | Oxacan Dwarf Boa | II | |
| *Liasis boa* | Ringed Python | II | |
| *Liasis boeleni* | Boelen's Python | II | |
| *Liasis childreni* | Children's Python | II | |
| *Liasis fuscus* | Australian Water Python | II | |
| *Liasis mackloti* | Macklot's Python | II | |
| *Liasis olivaceus* | Olive Python | II | |
| *Liasis papuanus* | Papuan Python | II | |
| *Liasis perthensis* | Perth Pygmy Python | II | |
| *Loxocemus bicolor* | Mexican Burrowing Python | II | |
| *Morelia amethistina* | Amethystine Python | II | |
| *Morelia carinata* | Keeled Scaled Python | II | |
| *Morelia oenpelliensis* | Oenpelli Python | II | |
| *Morelia spilota* | Carpet Python | II | |
| *Python anchietae* | Angola Python | II | |
| *Python curtus* | Blood Python | II | |
| *Python molurus* | Asiatic Rock Python | II | |
| *Python molurus molurus* | Indian Python | I | E |
| *Python regius* | Ball Python | II | |
| *Python reticulatus* | Reticulated Python | II | |
| *Python sebae* | Rock Python | II | |
| *Python timoriensis* | Timor Python | II | |
| *Sanzinia madagascariensis* | Madagascar Tree Boa | I | |
| *Trachyboa boulengeri* | Northern Eyelash Boa | II | |
| *Trachyboa gularis* | Southern Eyelash Boa | II | |
| *Tropidophis battersbyi* | Battersby's Dwarf Boa | II | |
| *Tropidophis canus* | Great Inagua Island Dwarf Boa | II | |
| *Tropidophis caymanensis* | Cayman Island Dwarf Boa | II | |
| *Tropidophis feicki* | Feick's Dwarf Boa | II | |
| *Tropidophis greenwayi* | Ambergris Cay Dwarf Boa | II | |
| *Tropidophis haetianus* | Haitian Dwarf Boa | II | |
| *Tropidophis maculatus* | Spotted Dwarf Boa | II | |
| *Tropidophis mela nurus* | Navassa Island Black-tailed Dwarf Boa | II | |
| *Tropidophis nigriventris* | Black-bellied Dwarf Boa | II | |

|  |  | CITES | E/T |
|---|---|---|---|
| *Tropidophis pardalis* | Spotted Dwarf Boa | II | |
| *Tropidophis paucisquamis* | Brazilian Dwarf Boa | II | |
| *Tropidophis pilsbryi* | Pilsbry's Dwarf Boa | II | |
| *Tropidophis semicinctus* | Banded Dwarf Boa | II | |
| *Tropidophis taczanowski* | Taczanowski's Dwarf Boa | II | |
| *Tropidophis wrighti* | Wright's Dwarf Boa | II | |
| *Ungaliophis panamensis* | Panamanian Dwarf Boa | II | |
| *Xenoboa cropanii* | Cropan's Boa | II | |

## Chamaeleonidae

| | | | |
|---|---|---|---|
| *Bradypodion* (all species) | dwarf chameleons | II | |
| *Chamaeleo* (all species) | chameleons | II | |

## Colubridae

| | | | |
|---|---|---|---|
| *Atretium schistosum* | Olive Keelback Water Snake (India) | III | |
| *Cerberus rhynchops* | Dog-faced Water Snake (India) | III | |
| *Clelia clelia* | Mussurana Snake | II | |
| *Cyclagras gigas* | South American False Water Cobra | II | |
| *Drymarchon corais couperi* | Eastern Indigo Snake | | T |
| *Liophis ornatus* | Maria Island Snake | | E |
| *Nerodia clarkii taeniata* | Atlantic Salt Marsh Snake | | T |
| *Nerodia paucimaculata* | Concho Water Snake | | T |
| *Ptyas mucosus* | Oriental Rat Snake | II | |
| *Thamnophis sirtalis tetrataenia* | San Francisco Garter Snake | | E |
| *Thamnophis gigas* | Giant Garter Snake | | T |
| *Xenochrophis piscator* | Checkered Keelback Water Snake (India) | III | |

## Cordylidae

| | | | |
|---|---|---|---|
| *Cordylus* (all species) | girdled lizards | II | |
| *Pseudocordylus* (all species) | false girdled lizards | II | |

## Crotalidae

| | | | |
|---|---|---|---|
| *Agkistrodon bilineatus* | Cantil (Honduras) | III | |
| *Bothrops asper* | Terciopelo (Honduras) | III | |
| *Bothrops nasutus* | Rhinoceros Lancehead (Honduras) | III | |
| *Bothrops nummifer* | Jumping Pit Viper (Honduras) | III | |
| *Bothrops ophryomegas* | Slender Hognosed Pit Viper (Honduras) | III | |
| *Bothrops schlegelii* | Eyelash Palm Pit Viper (Honduras) | III | |
| *Crotalus durissus* | Cascabel Rattlesnake (Honduras) | III | |
| *Crotalus unicolor* | Aruba Island Rattlesnake | | T |
| *Crotalus willardi obscurus* | New Mexican Ridgenosed Rattlesnake | | T |

## Elachistodontinae

| | | | |
|---|---|---|---|
| *Elachistodon westermanni* | Indian Egg-eating Snake | II | |

## Elapidae

| | | | |
|---|---|---|---|
| *Hoplocephalus bungaroides* | Broad-headed Snake | II | |
| *Micrurus diastema* | Atlanta Coral Snake (Honduras) | III | |
| *Micrurus nigrocinctus* | Black-banded Coral Snake (Honduras) | III | |
| *Naja naja* | Indian Cobra | II | |
| *Ophiophagus hannah* | King Cobra | II | |

|  |  | CITES | E/T |
|---|---|---|---|
| **Gekkonidae** | | | |
| *Cyrtodactylus serpensinsula* | Serpent Island Gecko | II | T |
| *Phelsuma* (all species) | day geckos | II | |
| *Phelsuma edwardnewtoni* | Newton Day Gecko | II | E |
| *Phelsuma guentheri* | Round Island Day Gecko | II | E |
| *Sphaerodactylus micropithecus* | Monito Gecko | | E |
| **Helodermatidae** | | | |
| *Heloderma horridum* | Mexican Beaded Lizard | II | |
| *Heloderma suspectum* | Gila Monster | II | |
| **Iguanidae** | | | |
| *Amblyrhynchus cristatus* | Galapagos Marine Iguana | II | |
| *Anolis roosevelti* | Culebra Giant Anole | | E |
| *Brachylophus fasciatus* | Fiji Banded Iguana | I | E |
| *Brachylophus vitiensis* | Fiji Crested Iguana | I | E |
| *Conolophus pallidus* | Barrington Island Land Lizard | II | E |
| *Conolophus subcristatus* | Galapagos Land Iguana | II | |
| *Cyclura* (all except as listed below) | ground iguanas | I | |
| *Cyclura carinata bartschi* | Mayaguana Iguana | I | T |
| *Cyclura carinata carinata* | Turks and Caicos Iguana | I | T |
| *Cyclura collei* | Jamaican Iguana | I | E |
| *Cyclura cychlura cychlura* | Andros Island Ground Iguana | I | T |
| *Cyclura cychlura figginsi* | Exuma Island Iguana | I | T |
| *Cyclura cychlura inornata* | Allen's Cay Iguana | I | T |
| *Cyclura nubila caymanensis* | Cayman Brac Ground Iguana | I | T |
| *Cyclura nubila lewisi* | Grand Cayman Ground Iguana | I | E |
| *Cyclura nubila nubila* | Cuban Ground Iguana (except Puerto Rico) | I | T |
| *Cyclura pinguis* | Anegada Ground Iguana | I | E |
| *Cyclura rileyi nuchalis* | Acklins Ground Iguana | I | T |
| *Cyclura rileyi rileyi* | Watling Island Ground Iguana | I | E |
| *Cyclura rileyi cristata* | White Cay Ground Iguana | I | T |
| *Cyclura stejnegeri* | Mona Ground Iguana | I | T |
| *Gambelia silus* | Blunt-nosed Leopard Lizard | | E |
| *Iguana delicatissima* | Antillean Iguana | II | |
| *Iguana iguana* | Green Iguana | II | |
| *Phrynosoma coronatum* | Coastal Horned Lizard | II | |
| *Phrynosoma coronatum blainvillei* | San Diego Horned Lizard | II | |
| *Sauromalus varius* | San Esteban Island Chuckwalla | I | E |
| *Uma inornata* | Coachella Valley Fringe-toed Lizard | | T |
| **Lacertidae** | | | |
| *Gallotia simonyi simonyi* | Hierro Giant Lizard | I | E |
| *Podarcis lilfordi* | Lilford's Wall Lizard | II | |
| *Podarcis pityusensis* | Ibiza Wall Lizard | II | T |
| **Scincidae** | | | |
| *Corucia zebrata* | Prehensile-tailed Skink | II | |
| *Eumeces egregius lividus* | Bluetail Mole Skink | | T |
| *Leiolopisma telfairi* | Round Island Skink | | T |
| *Neoseps reynoldsi* | Sand Skink | | T |

|  |  | CITES | E/T |
|---|---|---|---|
| **Teiidae** | | | |
| *Ameiva polops* | St. Croix Ground Lizard | | E |
| *Cnemidophorus hyperythrus* | Orange-throated Whiptail Lizard | II | |
| *Cnemidophorus vanzoi* | Maria Island Ground Lizard | | E |
| *Crocodilurus lacertinus* | Crocodile Tegu | II | |
| *Tupinambis rufescens* | Red Tegu | II | |
| *Tupinambis teguixin* | Black Tegu | II | |
| **Varanidae** | | | |
| *Varanus* (all except as listed below) | Monitor Lizards | II | |
| *Varanus bengalensis* | Bengal Monitor | I | E |
| *Varanus flavescens* | Yellow Monitor | I | E |
| *Varanus griseus* | Desert Monitor | I | E |
| *Varanus komodoensis* | Komodo Island Monitor | I | E |
| **Viperidae** | | | |
| *Vipera latifii* | Lar Valley Viper | | E |
| *Vipera russellii* | Russell's Viper (India) | III | |
| *Vipera ursinii* | Orsini's Viper (except USSR) | I | |
| *Vipera wagneri* | Wagner's Viper | II | |
| **Xantusiidae** | | | |
| *Xantusia riversiana* | Island Night Lizard | | T |
| **Xenosauridae** | | | |
| *Shinisaurus crocodilurus* | Chinese Crocodile Lizard | II | |
| **Testudinata—Chelidae** | | | |
| *Phrynops hogei* | Brazilian Sideneck Turtle | | E |
| *Pseudemydura umbrina* | Short-necked Swamp Turtle | I | E |
| **Cheloniidae** | | | |
| *Caretta caretta* | Loggerhead Turtle | I | T |
| *Chelonia depressa* | Flatback Turtle | I | |
| *Chelonia mydas* | Green Turtle | I | T |
| *Chelonia mydas* | Green Turtle | I | E |
| (breeding colony populations in Florida and on Pacific coast of Mexico) | | | |
| *Eretmochelys imbricata* | Hawksbill Turtle | I | E |
| *Lepidochelys kempi* | Kemp's Ridley Turtle | I | E |
| *Lepidochelys olivacea* | Olive Ridley Turtle | I | T |
| *Lepidochelys olivacea* | Olive Ridley Turtle | I | E |
| (breeding populations on Pacific coast of Mexico) | | | |
| **Dermatemydidae** | | | |
| *Dermatemys mawii* | Central American River Turtle | II | E |
| **Dermochelyidae** | | | |
| *Dermochelys coriacea* | Leatherback Turtle | I | E |
| **Emydidae** | | | |
| *Batagur baska* | River Terrapin | I | E |
| *Clemmys insculpta* | Wood Turtle | II | |
| *Clemmys muhlenbergi* | Bog Turtle | I | |
| *Geoclemys hamiltonii* | Spotted Pond Turtle | I | E |
| *Graptemys flavimaculata* | Yellow-blotched Map Turtle | | T |

**61**

|  |  | CITES | E/T |
|---|---|---|---|
| *Graptemys oculifera* | Ringed Map Turtle |  | T |
| *Kachuga tecta tecta tecta* | Indian Sawback Turtle | I | E |
| *Melanochelys tricarinata* | Three-keeled Asian Turtle | I | E |
| *Morenia ocellata* | Burmese Peacock Turtle | I | E |
| *Pseudemys alamabensis* | Alabama Redbelly Turtle |  | E |
| *seudemys rubriventris bangsi* | Plymouth Redbelly Turtle |  | E |
| *Terrapene coahuila* | Aquatic Box Turtle | I | E |
| *Trachemys scripta callirostris* | South American Red-lined Turtle |  | E |
| *Trachemys stejnegeri* | Malonei Inagua Island Turtle |  | E |
| *Trachemys terrapen* | Cat Island Turtle |  | E |

(Cat Island in the Bahamas)

## Kinosternidae

| | | | |
|---|---|---|---|
| *Sternotherus depressus* | Flattened Musk Turtle |  | T |

(Black Warrior River system upstream from Bankhead Dam)

## Pelomedusidae

| | | | |
|---|---|---|---|
| *Erymnochelys madagascuriensis* | Madagascar Turtle | II | |
| *Peltocephalus dumeriliana* | Big-headed Amazon River Turtle | II | |
| *Pelusios adansonii* | Adanson's Hinged Terrapin (Ghana) | III | |
| *Pelusios castaneus* | Brown Hinged Terrapin (Ghana) | III | |
| *Pelusios gabonensis* | Gabon Hinged Terrapin (Ghana) | III | |
| *Pelusios niger* | Black Hinged Terrapin (Ghana) | III | |
| *Podocnemis erythrocephala* | South American Turtle | II | |
| *Podocnemis expansa* | Arrau Turtle | II | E |
| *Podocnemis lewyana* | South American Turtle | II | |
| *Podocnemis sextuberculata* | South American Turtle | II | |
| *Podocnemis unifilis* | Yellow-spotted Amazon Turtle | II | E |
| *Podocnemis vogli* | Sabanera | II | |

## Pleurodira

| | | | |
|---|---|---|---|
| *Pelomedusa subrufa* | Helmeted Terrapin (Ghana) | III | |

## Testudinidae

| | | | |
|---|---|---|---|
| *Agrionemys horsfeldi* | Four-toed Tortoise | II | |
| *Chersinaa angulata* | Bowsprit Tortoise | II | |
| *Geochelone carbonaria* | Red-legged Tortoise | II | |
| *Geochelone chilensis* | Argentine Tortoise | II | |
| *Geochelone denticulata* | Yellow-footed Tortoise | II | |
| *Geochelone elegans* | Indian Star Tortoise | II | |
| *Geochelone elephantopus* | Galapagos Tortoise | I | E |
| *Geochelone gigantea* | Aldabra Giant Tortoise | II | |
| *Geochelone pardalis* | Leopard Tortoise | II | |
| *Geochelone platynota* | Burmese Tortoise | II | |
| *Geochelone radiata* | Radiated Tortoise | I | E |
| *Geochelone sulcata* | Spurred Tortoise | II | |
| *Geochelone yniphora* | Angulated Tortoise | I | E |
| *Gopherus agassizii* | Desert Tortoise | II | T |
| *Gopherus berlandieri* | Texas Tortoise | II | |
| *Gopherus flavomarginatus* | Bolson Tortoise | I | E |
| *Gopherus polyphemus* | Gopher Tortoise | II | T |

(west of Mobile and Tombigbee Rivers in Alabama, Mississippi, Louisiana)

|  |  | CITES | E/T |
|---|---|---|---|
| *Homopus areolatus* | Parrot-beaked Tortoise | II | |
| *Homopus boulengeri* | Donner-weer Tortoise | II | |
| *Homopus femoralis* | Karroo Tortoise | II | |
| *Homopus signatus* | Speckled Tortoise | II | |
| *Indotestudo elongata* | Yellow-headed Tortoise | II | |
| *Indotestudo travancoricae* | Travancore Tortoise | II | |
| *Kinixys belliana* | Bell's Hinged Tortoise | II | |
| *Kinixys erosa* | Schweigger's Hingeback Tortoise | II | |
| *Kinixys homeana* | Home's Hinged Tortoise | II | |
| *Malacochersus tornieri* | Pancake Tortoise | II | |
| *Manouria emys* | Brown Tortoise | II | |
| *Manouria impressa* | Indo-Chinese Tortoise | II | |
| *Psammobates geometricus* | Geometric Turtle | I | E |
| *Psammobates oculifer* | Serrated Geometric Tortoise | II | |
| *Psammobates tentorius* | Knobby Geometric Tortoise | II | |
| *Pseudotestudo kleinmanni* | Egyptian Tortoise | II | |
| *Pyxis arachnoides* | Spider Tortoise | II | |
| *Pyxis planicauda* | Flatback Spider Tortoise | II | |
| *Testudo gracea* | Spur-thighed Mediterranean Tortoise | II | |
| *Testudo hermanni* | Greek Tortoise | II | |
| *Testudo marginata* | Margined Tortoise | II | |

## Trionychidae

|  |  | CITES | E/T |
|---|---|---|---|
| *Lissemys punctata punctata* | Indian Flap-shell Tortoise | I | |
| *Trionyx ater* | Cuatro Cienegas Softshell Turtle | I | E |
| *Trionyx gangeticus* | Indian Softshell Turtle | I | E |
| *Trionyx hurum* | Peacock Softshell Turtle | I | E |
| *Trionyx nigricans* | Black Softshell Turtle | I | E |
| *Trionyx triunguis* | Three-clawed Turtle | III | |

> *"A reverence for original landscape is one of the humanities. It was the first humanity. Reckoned in terms of human nerves and juices, there is no difference in the value of a work of art and a work of nature."*
> —Archie Carr, *Ulendo*, 1964

# Notes